HISTORIC HATCHVILLE

Horse and Farm Country on Cape Cod

LES GARRICK

Forewords by Dale Hatch & Robert Alan Hatch

The History PRESS

Published by The History Press
Charleston, SC 29403
www.historypress.net

Front cover, bottom: Ninth hole at Cape Cod Country Club. *Courtesy of Friel Golf Management.* *Back cover, top*: Extended family of Ebenezer Amasa Hatch (standing, fifth from left) and his wife, Adelia (standing, far right), with their son Allen. *Courtesy of Jeanette Davis*; *bottom*: Coonamessett Inn at the end of its horseshoe-shaped driveway, circa 1945. *Courtesy of Gareth Jennings.*

First published 2014

Manufactured in the United States

ISBN 978.1.62619.511.0

Library of Congress CIP data applied for.

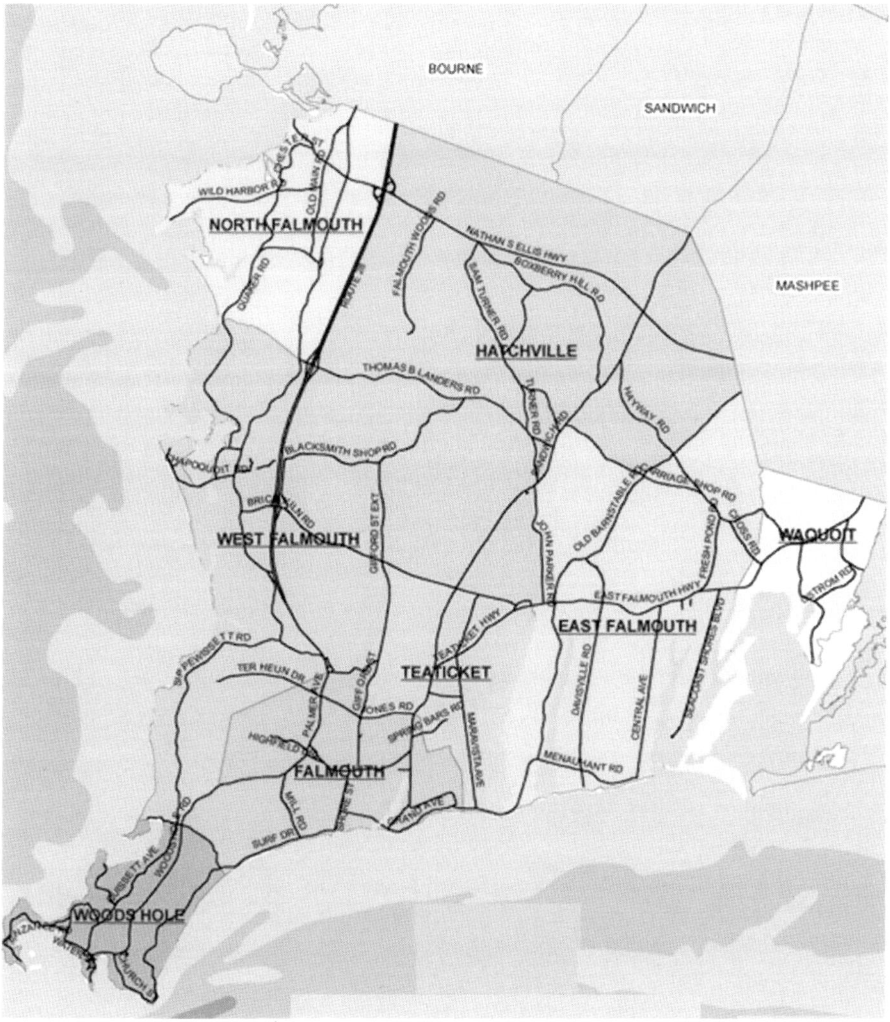

Eight Villages of Falmouth, Massachusetts. *Courtesy Falmouth Geographic Information System (GIS).*

CONTENTS

FOREWORD

The Hatch family owes much to Les Garrick for researching and writing *Historic Hatchville: Horse and Farm Country on Cape Cod*. It is the area where our Hatch ancestors settled after immigrating to escape the tyrannical monarch Charles I of England for religious freedom and liberty. It has been very interesting to read about the unique history of Hatchville and how it developed over many years through the words of Mr. Garrick.

Had I realized while attending history classes growing up that my Hatch ancestors were a part of American history, the classes in school and college would have had more meaning. I just did not get the connection until researching and writing the first of my three books published on the Hatch family. Our ancestors were early pioneers of the Puritan colonies in Massachusetts as early as 1634 or 1635 and, therefore, were among the early founders of this great nation of America.

Four generations of my direct Hatch line lived in the Falmouth/Barnstable area for almost one hundred years before moving on to Tolland, Connecticut. My eighth-great-grandparents Jonathan Hatch and Sarah Rowley were married on April 11, 1646, in Barnstable by Reverend John Lothrop (1584–1653). He was an English Anglican clergyman who became a Congregationalist minister. As a result of preaching his Puritan reform beliefs, Reverend Lothrop and members of his congregation were thrown into prison for two years. They had the options of denouncing their beliefs, being executed or moving to the New World. They chose to immigrate to New England. Lothrop founded Barnstable, Massachusetts, in 1639.

The Hatches were very religious people. In writing his last will and testament, dated September 15, 1705, Jonathan Hatch expressed some of his religious feelings and beliefs, stating he believed in God, life on Earth was transitory and there was life after death. It was his desire to live with God, whom he acknowledged as our creator. He believed in the resurrection of the dead through Jesus Christ, and he recognized and expressed gratitude to God as the source of his earthly possessions and property. Jonathan Hatch was honored and respected by the community where he had lived for the last fifty years of his life. He died at the age of eighty-five in Falmouth, Massachusetts, and is buried in the Falmouth Old Burying Ground.

One unusual and interesting story by Clarence J. Anderson (1912–2004) of Falmouth reads as follows:

> *Barnabas Hatch was born in Falmouth 8 December 1751. It is said that he married a near relative, which accounted for seven out of nine children being so small. They were less than four feet in height and four of them were so low of stature as to stand under a common door latch, but they lived to be quite aged and well educated and much respected. They were called the "The Little Hatches of Falmouth."*

Ancestors of today's Hatch family were very patriotic and lovers of freedom and liberty. In my research, ninety-one Hatch soldiers were in the American Revolutionary War, including my direct ancestors Nathaniel Hatch and his son Captain Jeremiah Hatch. After his father died in the war, Jeremiah joined up as a youth and was a fifer. Eleven of my direct fourth-, fifth- or sixth-great-grandfathers served as soldiers under General George Washington.

Three different Hatch lines (Captain Jeremiah Hatch, Ira Stearns Hatch and Jacob Hatch) joined the Church of Jesus Christ of Latter-Day Saints (Mormons) in the 1830s. Three generations of my ancestors in Addison County, Vermont, joined this church and moved to Nauvoo, Illinois, and then westward with the Mormon pioneers. This is why there are so many Hatches living in Utah, Idaho and Arizona today.

Jeremiah Hatch (1823–1903) spearheaded the formation of the Hatch Genealogical Society of Salt Lake City, Utah, in the 1890s. The research of this organization resulted in the publication of two volumes of a book entitled *Genealogy & History of the Hatch Family: Descendants of Thomas & Grace Hatch of Dorchester, Yarmouth, & Barnstable, Massachusetts* published from 1925 to 1930.

My genealogy, down to my great-grandfather, is as follows: Thomas, Jonathan, Benjamin, John, Nathaniel, Captain Jeremiah, Hezekiah, Jeremiah and Alvah Alexander Hatch.

Again, thanks to Les Garrick for writing this book and preserving the history of this historic area of Cape Cod.

Dale Hatch

FOREWORD

The same year I joined the Department of History at the University of Florida as a young assistant professor, Les Garrick and his colleagues were publishing important studies on the social behavior of that rowdy reptile known as the Florida gator. Coming from Wisconsin and the land of Bucky Badger, it took me a while to appreciate the scaly creatures that occasionally blocked traffic at the edge of campus. Those studies, I learned through my work with the Florida Audubon Society and the Sierra Club, helped conservation biologists understand the social structure of still-wild populations—in other words, which gators not to harvest for skins and steaks. Careful research—in science or history—helps inform our decisions.

Now retired from the University of Florida, I can devote more time to my Hatch heritage. And it is with great interest that I learned that Les was writing a book about the Hatches in Hatchville, a village of Falmouth, Massachusetts, on Cape Cod. Thomas Hatch, who arrived at Situate, Massachusetts, in 1638, is my oldest American ancestor. As it happens, two of the earliest Hatch settlers in colonial America were named Thomas Hatch. Here I confess that my Thomas did not go to Yarmouth and Barnstable, as did the other Thomas, from whose ancestors' name the name Hatchville derived. Instead, after several generations, part of my Thomas's lineage left Scituate for Attleboro, which would eventually become famous for jewelry and manufacturing—not the Native American's farmed wilderness.

Within three generations, change came to Attleboro. While life in Scituate involved farming and living off the ocean, Attleboro prompted occupational shifts from farming to trade, metalworking and commerce. Across some six generations in Massachusetts, these Hatch ancestors participated in the French and Indian War and the American Revolution, notably several tours of duty by Nathan Hatch, whose story ties together the westward movement of the family across four generations—from Massachusetts to Vermont, New York and Wisconsin. Nathan's brother became a family legend. Israel Hatch owned a number of taverns, some associated with his stagecoach line that linked Providence and Boston. His first effort was building the Hatch Tavern in Attleboro. Large and stately, it was a travel hub that drew such notable figures as George Washington and Thomas Jefferson. Although he served only briefly in the Revolution, Israel also owned establishments in Boston, including the White Horse Tavern, that were frequented by colonial activists and Free Masons. Israel also filed one of the first U.S. patents, one of them for a monorail train.

Nathan and his family eventually moved briefly to Halifax, Vermont, and following the birth of his several children, they pushed westward to Chautauqua County, New York. Although the family put down roots—with several children attending Fredonia Academy—around 1843, the enduring lure of open space and free land again drew them west to the Wisconsin Territory to clear land near Brookfield and then in Milwaukee County. Nathan died in 1847, in his ninetieth year, the only Revolutionary soldier buried in Waukesha County.

The Hatch family prospered in Brookfield until the Civil War, which changed everything. Edmund, the justice of the peace there, saw his three sons—Nathan Holbrook, Alanson and Leander Lycurgus—go to war. Twice wounded and permanently maimed, Leander was the sole survivor. His requests for a medical pension were denied annually, and he was buried without a headstone. At the end of the war, my great-great-grandfather Alanson died at the age of thirty-two. But here I can note one piece of good fortune. Thanks to a brief furlough from the Milwaukee Military Hospital (and timely favor from his wife's fertility clock), my great-grandfather Alanson Abbott Hatch was conceived just months before his father, Alanson, died of tuberculosis.

After the Civil War, Alanson's wife, Phoebe, married James Stanley, a wounded Civil War veteran. The extended family remnants—with Phoebe's mother and children from their earlier marriages—continued west in search of farmland, moving to or through Iowa, Minnesota, the Dakota Territory and the Pacific Northwest. Like his uncle Hiram, my great-grandfather A.A. Hatch farmed the land, ran a hardware store and later became the sometime mayor of Curlew, Iowa (population perhaps two hundred, about that of Hatchville in 1880). With the third generation in Iowa, my parents moved across the border to Minnesota, where my father was soon drafted into World War II. Near the end of the war, serving as an army cook and helping protect the borders of Texas, my father returned home, and I became the first family baby boomer. Make no mistake, I find this story personally appealing, and in some ways, it is representative of a pattern shared by many other American families.

But some stories have broader appeal and more enduring implications. This book aims to present a historical fabric that includes a number of families and a variety of evolving themes. The narrative is engaging not because it is seamless but because it exposes unexpected twists. As the author admits, it is an unconventional history. I take this to be a virtue. But I hasten to add that it is written by a determined researcher who relied on primary sources. Six years of labor produced an informed record of how a postage stamp–sized section of broader Hatchville came to command a special sense of place: horse and farm country and abundant open space. From the outset, the author shows that this outcome was not predetermined. If several unlikely events had not unfolded, this same land could have been residential sprawl. I enjoyed learning why.

There is something here for everyone. What I personally found intriguing was that the Hatches were the dominant family in Falmouth's East End—in the Congregational church, in accumulating the best land and in promoting self-serving legislation in the Massachusetts General Court. They were so influential that twenty-five-year-old Republican Silas Hatch (1833–1919) was named postmaster in 1858 by the Democrat president Buchanan, a position he held for sixty-one years. Silas even set a record by serving as a Falmouth selectman for thirty-five consecutive years. As a historian, I appreciated the author's balanced treatment of the major protagonists in this narrative. It put the land-use story in context. I had much to learn.

As conservationists, the author and I understand the fragility of open space and how easily it can be appropriated for other uses. We also see clearly that the effects of climate change at the local level must be prudently monitored if adaptation strategies are to be formulated in a noncrisis mode. This book addresses issues while offering a good read. Based on engaging, unlikely and unchangeable historical events, it ties a local history to a much broader web of challenges and possibilities.

Robert Alan Hatch

PREFACE

This book is a history of three hundred years of land use and the people who lived in a small part of the much larger current inland village of Hatchville in Falmouth, Massachusetts, on southwestern Cape Cod (Barnstable County). Hatchville is in Falmouth's northeast sector; it abuts the townships of Bourne, Sandwich, and Mashpee.

Why should historic Hatchville, a patch of approximately five square miles, be any more remarkable than other neighborhoods in Falmouth when it isn't even a historic district?

One reason is that the eastern part of the village of Hatchville is different. It is part of a broad valley of glacial outwash that over thousands of years developed excellent loam, which makes farming successful. Another is that there is also plenty of water for crop irrigation and livestock. It is also special because of the wildness of the landscape: waving grasslands, shaded pine and oak woods, sloppy cranberry bogs, a scenic golf course that hugs the northeast of a 158-acre great pond, a gentle herring run, a holly sanctuary and inviting kettle ponds. Besides, there is very little automobile traffic.

It was the challenge of discovering how historic Hatchville became a semirural island containing over two thousand acres of conservation land and open space within a cosmopolitan, seaside resort town with a significant tourist-based economy that engaged me. My reward for years of research was uncovering a series of improbable events initiated by some otherwise unlikely people with big personalities during the last one hundred years that resulted in large chunks of otherwise developable real estate remaining largely intact to become that cherished open space.

But some of the natural beauty of historic Hatchville would have been erased without successful battles to restrict inappropriate residential development during the late-twentieth-century Cape Cod land boom. We salute and recognize those determined citizens of Falmouth. Ultimately, though, this is a story about land use and people of different means and purposes in which is embedded a mystery.

My window into the world of historic Hatchville opened not from my interest in the Hatch family, for whom this village is named, but from my curiosity in 2003 about one of the many horticultural achievements of Wilfrid Wheeler, who owned and operated Ashumet Farm in Hatchville from 1925 until his death in 1961. Wheeler is one of those big personalities mentioned previously. During the 1940s and 1950s, Wheeler helped energize the American holly (*Ilex opaca*) business in the United States. Although I enjoy strolling among his extant holly plantings, I really wanted to learn more about Wheeler's selection and cultivation of new varieties of beach plum (*Prunus maritima*), a native woody shrub of the coastal Atlantic region whose delectable jelly has a passionate following.

Prior to moving to Hatchville from Concord, Massachusetts, in 1919, Wilfrid Wheeler had served on the Massachusetts Board of Agriculture, and during that year, he was the first commissioner of the Department of Agriculture. But it was his relationship to the prominent and wealthy Crane family of Chicago, who were known worldwide for their plumbing fixtures and other fabricated metal products, that ultimately would change the direction of Wheeler's professional life and the future of Hatchville.

Through my focus on Wheeler, I learned about other members of the Crane family circle who participated in the improbable series of events that led to a model farm, then a resort and, ultimately, decades later, acres of conservation land and open space in Hatchville. As a result, in this narrative, I have given special attention to the Crane family, specifically Charles R. Crane, sons Richard T. Crane and John O. Crane and daughter Frances Anita Crane, as well as Crane's son-in-law, Robert W. Leatherbee; Leatherbee's brother Frederick K. Leatherbee; and, of course, to Wilfrid Wheeler.

As I began my journey, I found that several local historians had provided invaluable information about Hatchville in *The Book of Falmouth*, published

Iconic native beach plum (*Prunus maritima*) fruit from Cape Cod. *Photo by author.*

for the town's tricentennial in 1986, a date that is critical for this narrative.[1] For example, they showed on 1880 maps where people had lived a century earlier. For all their diligent work, I was surprised by the omission of an entry in *The Book of Falmouth* about the farms, farmers and agricultural innovators in Hatchville from 1940 to 1960 and beyond. Fortunately, I was able to obtain firsthand family histories from this cohort so that I could recount their pre– and post–World War experiences.

HATCHVILLE TODAY

Falmouth has eight villages. Hatchville encompasses 29 percent of Falmouth's land area. Just to be clear, in Massachusetts, a village must be part of an incorporated town. The town sets village boundaries, largely for administrative purposes, which means that these boundaries can and have changed. Early in Falmouth's development, Hatchville was equivalent to what I have called historic Hatchville because that is where the Hatch family prospered.

Later, confusion about Hatchville's boundaries arose because it is no longer a census-designated place and has lost its postal zip code, having been lumped into East Falmouth for over forty years. Some residents have suggested that the loss of the post office diminished the village's identity. Another factor of confusion might be that the village has no commons or retail business center. In the eighteenth century, there was a tavern and a blacksmith's shop on the Falmouth to Sandwich Road, but today, without a village center to get their attention, people drive rapidly along the west–east Route 151 or north–south on Sandwich Road without knowing they are in Hatchville.

Commerce in Hatchville—represented by suppliers of construction materials such as sand, gravel and cement; the capped town dump; landscapers; and a sawmill—is restricted to the west side along east–west Thomas Landers Road where the soil is poor. Near the western terminus of Thomas Landers Road on the Buzzards Bay Moraine is the new ice arena and Technology Park, which hosts diverse businesses, the Social Security Administration and a 260-foot-tall, privately owned wind turbine.

To the immediate north of Hatchville is the thirty-square-mile Massachusetts Military Reservation, once home to massive World War II mobilization. To the east is the Barnstable County (now Cape

Cod) Fairgrounds, where 4-H clubs, gardeners and equestrians gather. Massachusetts Audubon's Holly and Wildlife Sanctuary (formerly Wilfrid Wheeler's Ashumet Farm) is just north of the fairgrounds.[2] In the southern portion of the current village is the eighty-six-acre Jenkins Pond around which a residential community called Pinecrest arose in the late 1940s.

Remnants of historic Hatchville include the East End Meetinghouse, built in 1797, which, in 1821, became the Second Congregational Church of Falmouth and, in 1982, became the Falmouth Jewish Congregation, at the junction of Sandwich and Hatchville Roads; several old farmhouses; a repurposed 1930 cow barn on Hatchville Road once part of the New Alchemy Institute; a ninety-two-year-old dark orange ceramic silo on conservation land along the Coonamessett River; and an eighty-year-old golf course on the northeast side of Coonamessett Pond that was once part of a sprawling resort.[3]

Today, except for a convenience store with gasoline pumps at the intersection of Route 151 and Sandwich Road (known as Clausons Corner) and the Coonamessett Farm store on Hatchville Road, there is no market at which to buy milk. During the first half of the twentieth century, there was at least one functioning dairy farm in Hatchville to supply the neighbors, and before that, most families kept a cow.

Relatively recent businesses in the village include the now defunct Cape Cod Winery on Sandwich Road, which produced some fine vintages, and Hatchville Pottery on Boxberry Hill Road just west of the Hatchville Road intersection.

While not having a shopping area nearby can be inconvenient, especially in the summer when automobile traffic in Falmouth is often unbearable, the residents endure the tiresome trips and then gladly retreat along narrow, winding roads to their solitude among the pines next to their favorite pond.

One resident described the village as "horse and farm country."[4] Obviously, historic Hatchville evokes a unique sense of place, which could be manifest as the feeling of serenity that harkens back to a quieter time.

Consider what one early twentieth-century traveler said about historic Hatchville, some of which rings true today:

> *One passes through forests to get there and finds tokens of rather ancient culture around the waters of beautiful Coonemoesett* [Coonamessett Pond]. *There are cranberry bogs around the lake above the water level, with pumps for flooding, and other bogs follow down the natural grades of the outlet valley. Some of the farms show excellent culture and a variety of crops, including apple orchards well laden.*[5]

This narrative, which explores both why and how semirural historic Hatchville has persisted, follows a thread from subsistence farming to model farm and land sales, then to the resort and new farms and finally to residential development schemes and conservation land.

The book is in three parts. The theme of Part I is how an audacious plan by the Crane family circle to buy thousands of acres of contiguous land not only in Hatchville but also in the abutting townships of Bourne, Sandwich and Mashpee set the course for land use in historic Hatchville. Wilfrid Wheeler had a plan for this purchase: to establish a model farm in Hatchville and provide an opportunity for new farmers to learn modern techniques and later buy land to farm. To organize this effort, the family circle formed the Coonamessett Ranch Company, which was once considered mysterious. To appreciate the context for this purchase, I describe the early years in the East End and the advent of the village of Hatchville in 1858. The Coonamessett Ranch Company only bound the Crane family circle together for about eight years, until 1924. After that time, we follow each member in her/his own unique life trajectory. For example, Wilfrid Wheeler and Robert W. Leatherbee became important Hatchville farmers and Falmouth citizens.

Part II is about agrarian life near a large resort on Coonamessett Pond. It begins in 1925, when the Coonamessett Ranch Company abandoned the model farm and, in its place, began to develop a resort. Meanwhile, as the resort was growing, in 1935, the Coonamessett Ranch Company sold 8,745 acres (62 percent of its land) to the Commonwealth of Massachusetts for the expansion of training facilities for the Twenty-sixth Division of the Massachusetts National Guard on Cape Cod. The influx of military personnel and federal money to the region led to an expansion of the resort. Part II ends with vignettes from people who lived and worked near this sprawling resort.

Part III follows the challenges to the character of historic Hatchville. It starts with three separate sales of the Coonamessett Resort assets by John O. Crane and the estate of his sister, Frances Anita Crane: 1,562.1 acres in Hatchville for a wildlife management area, the six-hundred-seat Falmouth Playhouse to Sara Gordon and finally the resort and its buildings to the Clauson family. The Clausons operated the resort from 1957 to 1977. When Mrs. Clauson sold the resort, Cape Cod was already experiencing a building and population explosion. The next decade would be a make-or-break time for Falmouth: could the town organize and exclude oversized and inappropriate residential developments that would ruin the agrarian character of historic Hatchville?

The historic Hatchville of today will not be the same twenty years from now. Climate change, municipal uses of land, sale of the private golf course and pressure to build more residential housing each could disturb the serenity of this place. Historic Hatchville needs public use and monitoring of its open spaces to generate public involvement to chart its future.

Les Garrick
Woods Hole, Massachusetts
2014

ACKNOWLEDGEMENTS

This work has been a relatively long journey of exploration as I sought to uncover how historic Hatchville came to be the way it is by highlighting the people involved. Along the way, there were several critical moments: when Barbara Gill, the town archivist in Sandwich, unearthed a document about the Homestead Trust; when Harriet Needham showed me photographs of the Leatherbee clan; when Harvey G. Clauson Jr. gave me a map of the "Coonamessett Property"; and especially when Richard James Leatherbee (grandson of Robert Leatherbee) shared his family documents and photographs. Thank you, all. Those foregoing items truly helped me anchor this narrative, but without the assistance and kindness of many others, my task would have been much more difficult.

For help with research and/or providing documents in their care, I thank Jennifer Gaines and Susan Witzell, Woods Hole Historical Museum; Lisa Raymond, MBL-WHOI Library; Lucy Loomis, Sturgis Library; Jill Erickson, Falmouth Public Library; Margaret McCormick, Woods Hole Public Library; William Hough, the *Enterprise*; Carolyn Powers and Mary Sicchio, formerly at the Falmouth Historical Society; and Meg Costello, Falmouth Historical Society. For assistance and access to Falmouth, Massachusetts Town Records, I am extremely grateful to the Assessor's Office, the Planning Department and the town clerk. Special thanks goes to Bob Shea of the Falmouth Geographic Information System (GIS) for making available aerial images and the village map of Falmouth, Tom Stone for providing a simple Falmouth map and Martin L. Monk for video help. Without the assistance of Sue Young, I couldn't have navigated the Barnstable County Registry of Deeds.

I also thank the students of Lawrence Junior High School in Falmouth and their teacher Dr. Marian M. Desrosiers for sharing their Hatchville photographs and enthusiasm for a sense of place.

For use of memorabilia and photographs or for agreeing to be interviewed, I thank Eugene Allen, Louise Allen Bailey, Carol Baker, Bronwen Ballou and the Bradley family, Edward Chute, Nancy Williams Chute, George Clauson, Harvey G. Clauson Jr. (deceased), Elizabeth Clauson, Jeannette Davis, Henry Dick, Richard Dimmock, Harry Donald, Hollis Engley, Margaret Hopkins Freeman (deceased), Michael Freeman, Elizabeth Friel, Elizabeth Gladfelter, Anita Gunning, Louise Houle, Gareth Jennings, Fred Jensen, Greta Jensen Tomic (deceased), Henry Jensen, Donna Jewett, Rolf Kurt Kluever, Nellie Kurz, Jeri Lavender, Elias Lieberman, David Lillie (deceased), David Mauk, Ann McDermott, William Orsi, Margaret Russell, Beth Schwarzman (deceased), Daniel Small, Ronald Smolowitz, The 300 Committee, Edward and Barbara Towers and Richard Warren Wheeler. Also, I thank Paul Rifkin for his excellent video recording of some of these interviews.

For helpful comments on the book in draft form, I thank Jessie Gunnard, Susan Witzell, Leonard Miele and Nannette Drake Oldenbourg. Marilyn Rowland did the indexing. I am grateful for the encouragement offered by Patricia Gadsby and the patience and support of Rita Anne Garrick.

PART I

IMPROBABLE EVENTS AND AN AUDACIOUS PLAN

Chapter 1

FALMOUTH'S EAST END

According to a respected genealogy, four distinct branches of the Hatch line settled in the English colonies—two in the Plymouth Colony and one each in Maine and Virginia. All of them contributed to the spread of Hatches throughout North America. By 1880, there were 10,438 enumerated Hatches, with Massachusetts, New York and Maine having the most sizeable populations.

In Falmouth, Massachusetts, the Hatch clan was influential in the young town. Its influence, which included Congregational church deacons, began when Jonathan Hatch (1625–1710) helped settle Succanessett (as southwestern Cape Cod was called by the native people), before it became Falmouth.

In 1661, Jonathan Hatch and Isaac Robinson were the first European settlers in Succanessett. Apparently, the pair and about a dozen other men were not welcome in Barnstable because of their tolerance of the Quakers. Soon, as proprietors, group members divided the land among themselves. Perhaps they all met at Jonathan Hatch's public house, where much business was transacted. The first settlements were along the wooded rolling terrain near the moraine and some flatter areas in the region's western coastal section, extending from present-day Woods Hole north to West Falmouth and the plains around Falmouth village. The proprietors and especially Jonathan Hatch had good relations with the Wampanoag, and thus their settlement avoided the usual land disputes between colonists and the native tribes, which exploded into a take-no-prisoners war in 1675 to the west and north of Cape Cod.

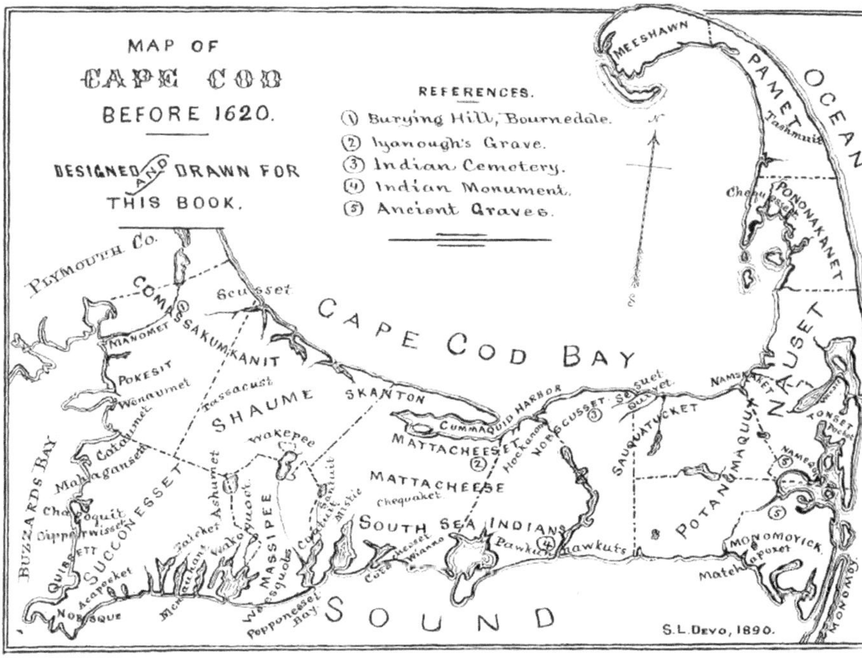

Boundaries of the Indian lands on Cape Cod before 1620. *From Simeon L. Deyo, History of Barnstable County, Massachusetts.*

Jonathan Hatch and his wife, Sarah Rowley, had eleven children. Their son Moses Hatch (1662–1747) was the first European child born in Succanessett; he later became a town administrator. During his fifty years in Falmouth, Jonathan Hatch became quite land rich. A generous man, prior to his death, he gave away all of his land to family members. Still, some of his children left the area and thus contributed to the spread of the resourceful Hatch line throughout North America.[6]

EASTWARD EXPANSION

In order to more easily reach the gristmill on Five Mile River, which was the eastern boundary of the original land purchase from the Wampanoag, the settlers constructed a wider road from Little Harbor in Woods Hole. Scattered settlements soon followed the new road to the east side of Succanessett. In the 1830s the river was called Dexter's River, when a woolen mill replaced the gristmill, and later still it was renamed the Coonamessett River.

Five years after the town was incorporated in 1686, Falmouth purchased land north to the Sandwich line, west to North Falmouth and east to the Mashpee line from the Wampanoag. In 1712, the lots, which ran northeastward from the moraine toward the new town boundaries, were laid out.

Besides the lure of land for a growing population, the East End had two major benefits for the new settlers: several feet of fine loamy soil for farming and freshwater ponds for irrigation and ice.[7] Although the East End is dotted with ponds, the main ones targeted for settlement were the 27-acre Deep Pond, 36-acre Crooked Pond and 158-acre Coonamessett Pond (often referred to as a lake), which is spring fed from below, as is the Coonamessett River that flows to Nantucket Sound. Contrary to misinformed opinion, there are no freshwater rivers from the north bringing water to Cape Cod; the only source of fresh water is precipitation, which averages about forty-eight inches per year. This condition is called a sole-source aquifer.

In this so-called new purchase, different Falmouth residents were allotted a hill (near the moraine) and/or a plain lot extending two to three miles in a northeast direction toward the new boundaries.[8] These lots were bought, sold, shared, inherited and likely aggregated to many acres. As a result, several families—Hatch, Turner and Robinson—were able to set deep roots in the East End.

Ebenezer Hatch (1708–1796), a direct descendant of founder Jonathan Hatch, was probably the first Hatch in the East End. He and his wife, Mercy Crocker, settled near Deep Pond and had nine children. One of his sons, Moses (1762–1855), and his grandson Sylvanus (1789–1855) established a family compound consisting of farmhouses, barns and other outbuildings north of Coonamessett Pond. The Turners settled on the southwest side of Coonamessett Pond, and the Robinsons owned land south of Coonamessett Pond adjacent to the Coonamessett River. Eventually, in the East End, the Hatches became more numerous and owned more valuable farmland than other families.

The East End farmers cleared the wooded land with axes, saws and brushfires and used oxen to pull stumps and move boulders. They obtained fresh water from the ponds and drilled wells to reach the deeper layers of ground water. They built half, three-quarter or in some cases full Cape clapboard covered houses roofed with wood shingles. When a farming family outgrew their basic house, additions were spliced on to the core dwelling.

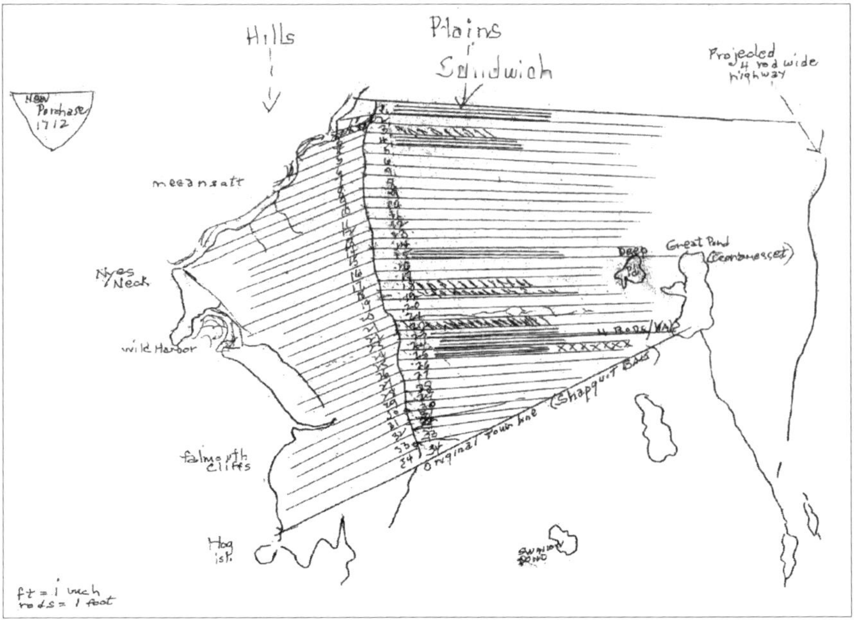

Layout of hill and plain lots in 1712 for Falmouth's eastward expansion. *From Svenning, "18th Century Deeds," 2001.*

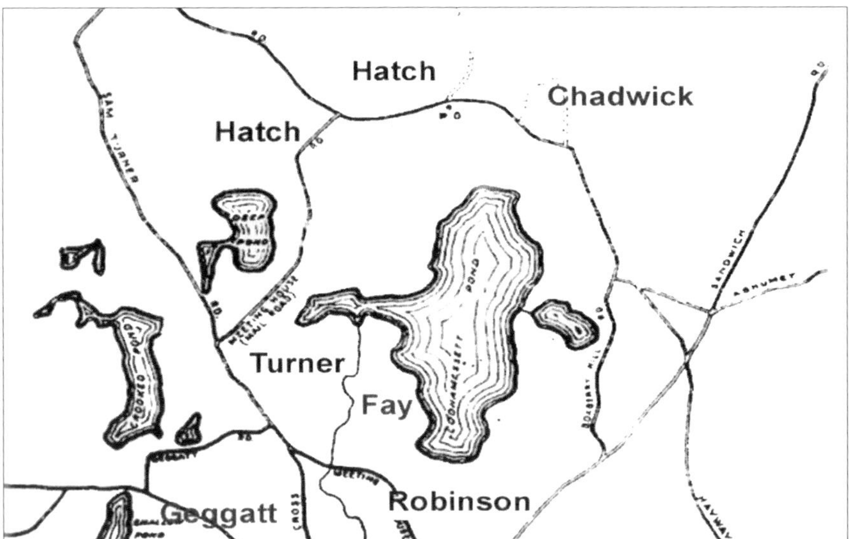

Major settler families of the East End of Falmouth, Massachusetts. *Annotated by author from a map in public domain.*

Team of quality cattle in circa 1900 Hatchville. *Courtesy of the Falmouth Historical Society, Falmouth, Massachusetts.*

The settlers farmed the rich loam, raising primarily corn. They kept a few cows and some chickens but not sheep like in the west of Falmouth. They established fruit and nut trees, harvested wild cranberries and caught herring migrating up river, which they ate and used for fertilizer. Roads were cut to provide access to the gristmill and to beaches on Buzzards Bay and Waquoit Bay to harvest salt hay for animal feed and kelp and eel grass to enrich the soil. The later track is still called Hayway Road and was once known as "Hatches Hayward."

There was a bounty on wolves, and it seems that the last wolf was killed on Cape Cod in Sandwich around 1837. By removing potential predators, farmers then had to deal with groundhogs, for which there was a bounty that extended into the mid-twentieth century.

EAST END MEETINGHOUSE

It was not long before the population of the East End grew sufficiently large to support its own house of worship and related cemetery. By a vote of the first congregation at the Falmouth Meetinghouse in 1795, the East End parish was established.

East End Meetinghouse (Second Congregational Church). *Photo by author.*

Congregationalists used buildings for civic meetings and worship. It was the state religion in Massachusetts, which excluded Catholics from entering the state on pain of death. Law dictated tax support for the church, especially for a pastor and his upkeep. Massachusetts was the last state to separate religion and government, doing so in 1833.

In keeping with the traditions of the Congregational Society, Ezekiel Robinson donated land for the new meetinghouse and burying ground, in which Jonathan Hatch, a Revolutionary War veteran and son of Ebenezer and Mercy Crocker, was the first adult buried. Ebenezer died soon after his son. In 1797, the meetinghouse was built on about an acre. The structure was built like a barn so as to not appear ostentatious, and it had a pitched roof with the long side of the building facing Sandwich Road.

In 1821, the East End parish organized as the Second Congregational Church of Falmouth. At that time, there were seventy-six church members, nine of whom were Hatches. Between 1821 and 1917, four Hatch men served as deacons. They (with their years of service) were Benjamin (1821–61), Sylvanus (1835–41), Benjamin H. (1868–92) and James T. (1893–1911). Four other Hatchville men also served as deacons.

In the first decades of the nineteenth century, unlike the town's coastal western villages of Woods Hole, Falmouth and West Falmouth, the inland East End did not directly garner the robust financial rewards of the flourishing seaside saltworks, coastal trading or whaling industry. According to the 1820 federal census, Falmouth's economy was roughly one-third each commerce, manufacturing and agriculture. The more agrarian East End religious community supported the meetinghouse the best way it could.

However, it wasn't until 1841 that the East End Meetinghouse was better supported. Businessman Shubael Lawrence owned a fulling mill (used to finish wool from the once abundant supply in Falmouth) on the Coonamessett River and land south of the East End Meetinghouse. After his father's death in 1841, Shubael Lawrence Jr., who also died in 1841, bequeathed $10,000 to establish a fund whose interest would, in the future, support a minister or religious teacher, provided that within two years of his (Lawrence Jr.'s) death, the congregation, at its own expense, rotate the meetinghouse so that its gable end was to the road, add a steeple, purchase a bell and keep the church in good repair. The Second Congregational Society, at a cost of $2,200, fulfilled these conditions. There was no heat in the meetinghouse until 1842.

HATCH FAMILIES

Hatchville's rolling landscape flattens out considerably to the north and west of Coonamessett Pond, yielding a vista, if planted in grain or corn, not unlike a field in Iowa. In its natural state, the community of plants and animals would be a type of sand plain grassland, like a prairie. Few dirt roads disrupted this landscape, just hundreds of acres of pasturage and corn. West of Sandwich Road, the land rises gradually to this plain (at ninety feet above sea level), which is close enough to Coonamessett Pond (at twenty to thirty feet above sea level) so that livestock would have access to fresh water via a culvert. Some of the Hatch clan settled along a road adjacent to the plain, called Depot Road, which later was called Boxberry Hill Road for the wintergreen plant that was found there. In this setting, some of the Hatches prospered during the mid- to late nineteenth century.

Silas Hatch (1833–1919) lived on this stretch of Depot Road. He was the most eminent of the clan because he served as a Falmouth selectman continuously for thirty-five years, retiring in 1911. This fully bearded gent

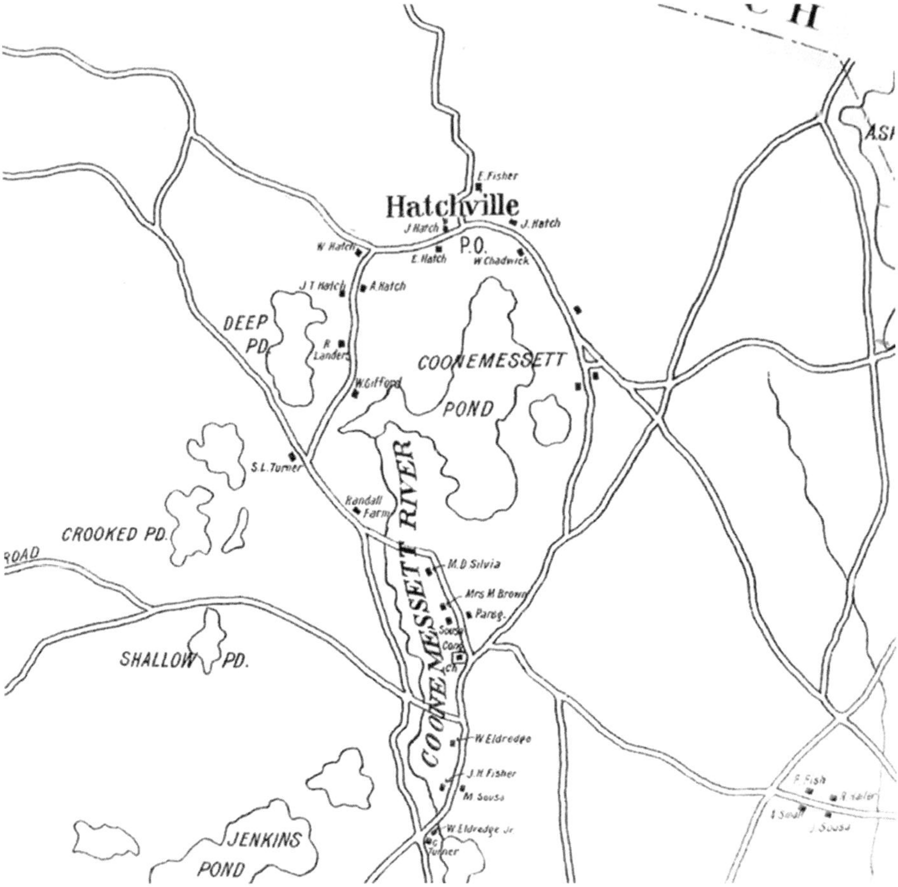

A 1910 Hatchville map. *From* Atlas of Barnstable County, Massachusetts.

was involved in many of the interesting and important mid-nineteenth and early twentieth centuries' events in Falmouth. Consider one case that happened during the end of his tenure: Selectman Silas Hatch and several other prominent citizens were asked in 1907 to mediate between a group of wealthy summer residents who had proposed seceding from Falmouth and creating a new town called "Nobska." The summer residents felt they were overtaxed to pay for town improvements. The threat was eventually withdrawn after it was agreed at the Falmouth Town Meeting to cut the budget and fund some pet projects of the shore dwellers.[9]

Silas and his four elder sisters (Adelia, Phebe, Clarissa and Chloe) were the children of Sylvanus Hatch (1789–1855), who was a church deacon

Silas Hatch, Hatchville postmaster and one of Falmouth's selectmen, in 1896. *Courtesy of the Falmouth Historical Society, Falmouth, Massachusetts.*

and a captain at the coast during the War of 1812, and Tirzah Hatch (1797–1887), the daughter of Isaac Hatch and Rhoda Robinson. Except for Chloe, who remained with Silas, the other sisters married and resided outside Falmouth. Their grandfather was Moses Hatch (1762–1855), and their great-grandfather was Ebenezer Hatch (1708–1796).

Silas was born on the Hatch farm on Depot Road in the East End in 1833. He had a twin who did not survive the year. Presumably, he was named for his late fraternal uncle Silas W. Hatch (1798–1830), who married Sarah Hamlin and farmed in Hatchville. Silas's two cousins from this marriage—Thatcher C., a merchant, and Moses, a businessman—both lived in cosmopolitan New Bedford, Massachusetts.

Silas's first wife, Henrietta Davis (1839–1883), who was disabled late in life with "softening of the brain," was the mother of Herbert C., Eugene P. and Cora M. His second wife was Mrs. Elizabeth M. (Bourne) Dillingham.

Getting to and exploiting the best seashore and marsh resources, especially salt hay, was important for the farmers. Led by Silas Hatch, this influential family set their sights on the east side of Waquoit Bay, about seven miles from their homestead. Even though they owned the right of way to

Hatchville De

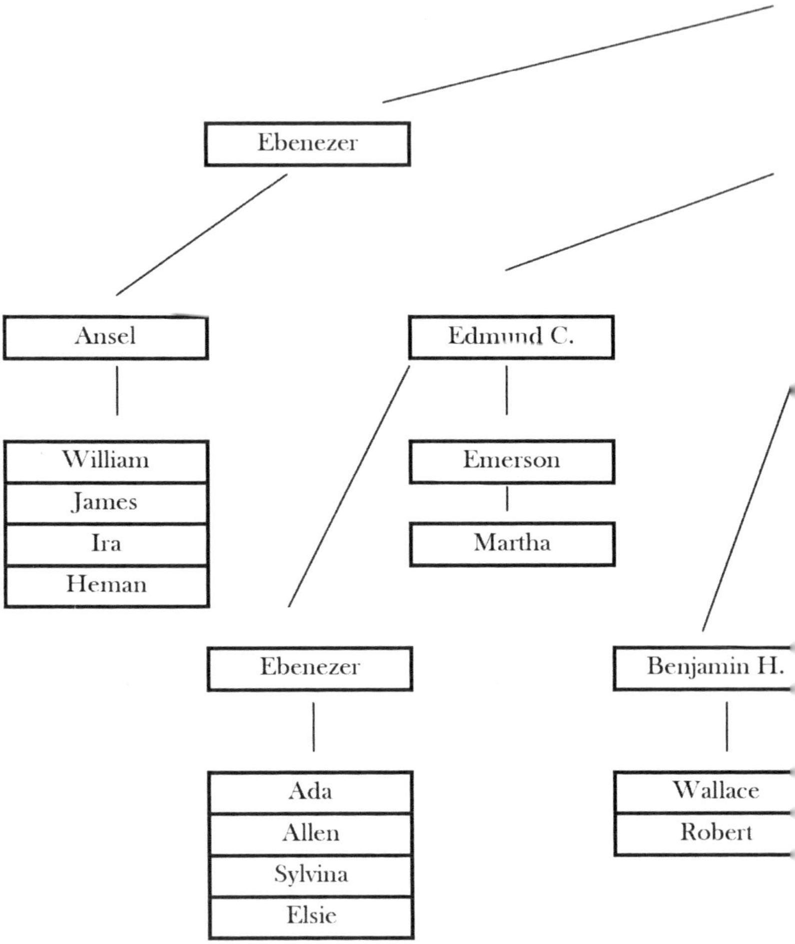

Bold names with birth and death dates are the direct male ancestors of Silas Hatch and his children. *Created by the author.*

dants of Ebenezer Hatch

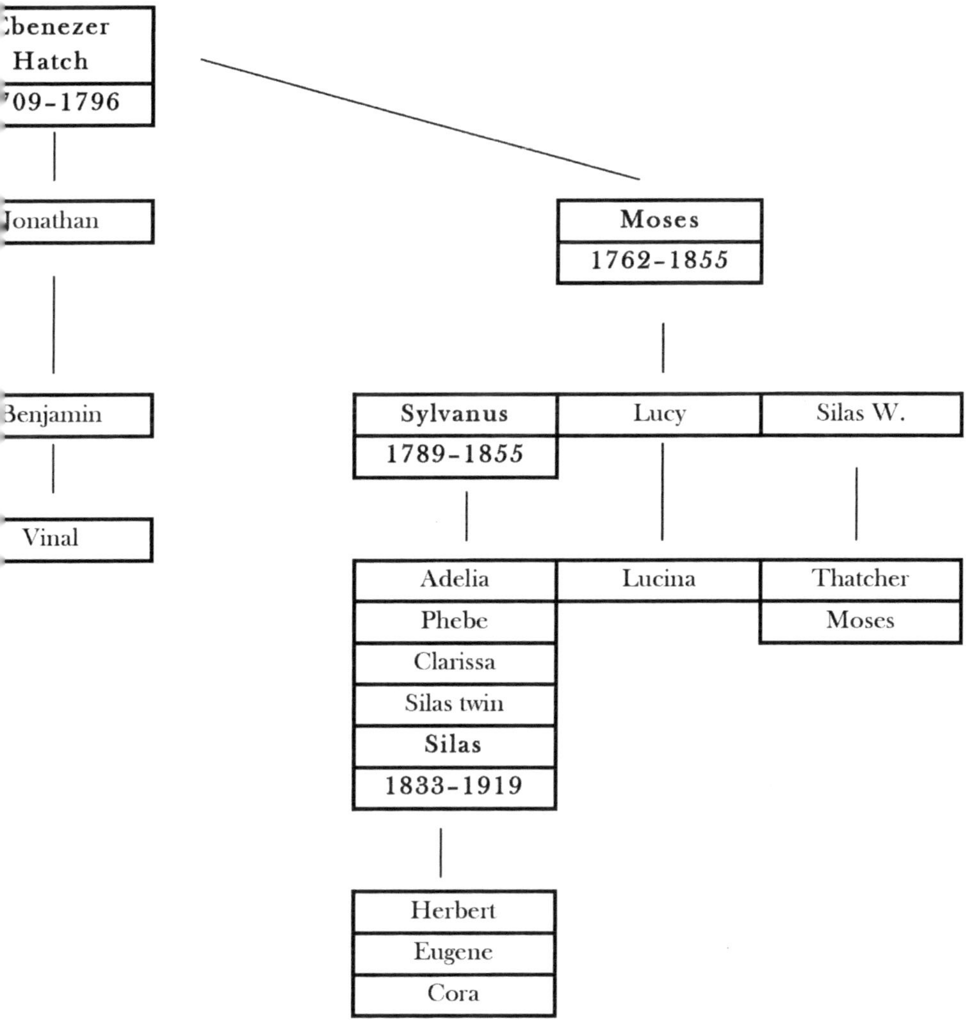

the eastern shore of Waquoit Bay, in 1857, they successfully petitioned the Massachusetts General Court to permit them to construct and maintain a special cart way across a creek.[10]

In 1858, President James Buchanan, a Democrat, appointed twenty-five-year-old Republican Silas Hatch as the postmaster of the small village of Hatchville, a position he held for sixty-one years. This presidential appointment certainly points to the political influence of the Hatch family. Most likely that is when Hatchville got its official name. Prior to 1858, John Robinson, the postmaster from East Falmouth, brought mail to Sunday worship at the East End Meetinghouse. The Hatchville Post Office, located in Hatch's home on Depot Road, was the village meeting place, as there was no general store. In 1883, Silas Hatch's compensation was $59.66, compared to the $822.00 earned by Falmouth's postmaster J.C. Burgess.[11]

According to the 1860 census, Hatches made up one-fourth of the 162 people among fourteen different families enumerated in Hatchville; there were no foreign-born residents then. These forty Hatches were all related to one another.

HERRING

Herring (alewives) were an important commodity along the Atlantic coast (they were salted and exported), a source of food (especially protein and fish oils) and fertilizer. In Hatchville, they could be caught in the shallows of the Coonamessett River in March or April when they swam up from the ocean to spawn in the Coonamessett Pond or when some of them returned to the sea later in the year. Native people are said to have maintained the connection between Coonamessett Pond (average depth nineteen feet) and the river. Later, in about 1847, in order to supply more water to the woolen mills along the Coonamessett River and to help the herring move up and down stream, the Town of Falmouth widened, deepened and lined with timber an existing channel between Coonamessett Pond and the river. For some reason, this channel is known as Dutchman's Ditch. It is still a good place to view migrating herring.

Herring could be very profitable, and some of the Hatch family used their political influence to attempt to succeed in this business in a novel way: Henry F. Hatch (a photographer), Robert H. Hatch, Benjamin H. Hatch and Isaac H. Hatch desired to own a herring run. They couldn't just buy an existing stream or river, which were part of the commons. The next best thing, they thought, was to dig a new two-mile-long channel on land they owned between Bourne's Pond on Nantucket Sound to fifty-six-acre Ashumet Pond in the most northeast corner of Hatchville. In March 1863, the Massachusetts General Court permitted them to form the East Falmouth Herring River Company.[12] Everything went according to plan, except the herring couldn't make it to Ashumet Pond; the ditch they dug is still visible south of Ashumet Pond.

Eventually, the Coonamessett River herring catch declined from overfishing and from the milldams and later cranberry bogs that diverted the flow. To remedy these problems, new herring regulations were imposed to give the herring preferred passage. Despite these regulations, conflicts over herring fishing became so dangerous that at the March 6, 1894 Falmouth Town Meeting, the three Herring Committee members were instructed to protect themselves from assault by carrying guns.

FIRES: DOMESTIC AND WILD

Fires have always been feared and were a threat to wooden dwellings with central hearths. To extinguish a house or barn fire, water was carried by bucket brigade from the ponds. In 1897, a limited fire department, which relied on volunteers, was funded. The first hose company was established in East Falmouth in 1906.

Uncontrollable forest fires threatened Hatchville, which was adjacent to thousands of acres of woodland and grassland, at any time of the year. James J. Hatch was the forest fire warden from 1879 to 1914, during which he witnessed many dangerous fires, which could only burn themselves out. In this climate, sun and wind will dry grasses in one hour and forest tinder in ten hours. Some of the wild fires were started by sparks from locomotives rolling south to Woods Hole west of the moraine and fanned by prevailing southwesterly winds, as happened in April 1900. In July of the same year, a human-caused fire spread through the dry woodlands of Sandwich and Bourne southeasterly into Hatchville, ultimately burning twenty-five square miles; the fire spared the Hatches' farmland. Severe wild fires of similar size also occurred in the 1920s and 1930s. A forest fire tower finally was built on the Buzzards Bay Moraine; even so, the risk of forest fire is still great. And despite all the close calls, Hatchville has never had a fire station.

CRANBERRY CULTIVATION AND PORTUGUESE IMMIGRATION

Wild cranberries were always part of the colonial diet, and mariners took them on their voyages to prevent scurvy. But as the techniques of cranberry cultivation spread in the mid- to late 1800s, acidic bogs, swamps and riverside land became more valuable, and cranberry bogs of about an acre in area were constructed. Their increasing value resulted in many bogs being bought and sold so that most of the Hatchville farmers also owned cranberry bogs. Typical of the new bogs was the large one of about six acres at the north end of Coonamessett Pond that was diked, irrigated and shared by different owners.

Retired sea captains like Charles Henry Turner, who lived on Sandwich Road after he returned to Hatchville, farmed cranberries along the Coonamessett River. He later moved to Falmouth village. James H. Fisher was another mariner who returned to his native Hatchville. James went to sea at age eighteen, sailed from 1856 to 1879, mainly out of New Bedford, and retired as a shipmaster. He probably used the shares from his whaling voyages to buy land and cranberry bogs. For example, in 1863, he bought property on Sandwich Road north of his parent's home on that road along the Coonamessett River that contained a cranberry bog, which later was maintained by his widow, Rebecca.

The biggest deals occurred in 1891 as commercial cranberry bog development exploded in Falmouth. At that time, according to assessors' records, Hatchville's cranberry bogs were valued at about $275 per acre while pasture was valued at $15 per acre and woodlots at much less. In that year, retired mariner James H. Fisher sold seventeen acres on Coonamessett Pond to wealthy Woods Hole landowner Joseph Story Fay, land Fisher had purchased in 1872. Fay, in 1891, also bought from Benjamin Hatch, Will Chadwick, Harry Turner and George Turner a total of fifty-four acres of upland, swamps and cranberry bogs at the south end of Coonamessett Pond. Fay's significant real estate purchase was the first large investment by someone not from Hatchville.[13]

Meanwhile, also in 1891, the Swift Brothers (Frederick, Howard and Russ), who purchased the holdings of a woolen company, began to convert twenty-five acres of swamps and millponds along the Coonamessett River into cranberry bogs. To work these cranberry bogs, the Swifts hired seasonal workers from New Bedford and Fall River who were immigrants from the Portuguese Azores and Cape Verde. In 1900, the Portuguese were 8

Above: Workers on the six-acre cranberry bog north of Coonamessett Pond. *Courtesy Harry Donald.*

Left: Joseph Story Fay, a Woods Hole summer resident, who in 1891 was the first outside land investor in Hatchville. *Courtesy Woods Hole Historical Museum Archives.*

percent of Falmouth's population of 3,500 and 20 percent of the town's 3,144 residents in 1910.[14] Commercial cranberry production was for many decades important for Hatchville and for this region's economy.[15]

Many of the immigrant Portuguese cranberry workers settled in Hatchville and East Falmouth and grew strawberries and other commercially important crops such as beets, potatoes, carrots and turnips on small plots of land.

PORTUGUESE LAND PURCHASES IN HATCHVILLE

The year 1891 was also noteworthy for the first purchase of Hatchville farmland by Azorean immigrants. Manuel Ignos and his wife, Margaret, came from Rhode Island and purchased their Hatchville farm. Manuel Dutra and his wife, Mary, also came to Hatchville in 1891. Mr. Dutra worked for the town of Falmouth, and he farmed on one hundred acres. Other Azoreans who became landowners in Hatchville around this time were the families of Manuel Enos, Frank Perry, Manuel Souza, Joseph Antone and Manuel Sylvia. Until these Portuguese arrived, immigration to Falmouth was rare except for the Irish workers who lived in Woods Hole and worked for the Pacific Guano Company there until it closed in 1889.[16]

By 1900, Hatchville's population was 250 (7 percent of the town's residents): twenty-eight were Hatches and 20 were foreign born, most of whom were from the Azores. The adult Portuguese were younger and had larger families than their established neighbors. By 1910, four immigrant families whose male heads either farmed or worked for a farmer lived along eastern Depot Road. Among their local compatriots, Manuel Sylvia operated a truck farm and Joseph Rose cultivated cranberries.

HATCHVILLE GRAMMAR SCHOOL

A single-room grammar school was built in Hatchville at the Ashumet and Depot Roads (Boxberry Hill Road) intersection in 1867. Silas Hatch was on the school committee and actually maintained the single-room schoolhouse for which in 1873 he was paid $4.00 and an additional $14.99 for painting it. In the 1870s, the school had a winter and a summer term, hiring separate teachers for each. Mr. Dillingham, a Hatch boarder, taught one of the

The 1898 class of Hatchville's one-room school. The teacher, Dorthea Collier, is standing in the last row on the far left. *Courtesy Jeanette Davis.*

sessions in 1870. Still, most of the teachers in the ten Falmouth schools at that time were unmarried females. The school superintendent remarked in the 1872–73 report that female teachers were preferred because they "excel in tact and skill in imparting instruction and in discipline."

Children of Azorean families made up more than one-third of the students in the class of 1898. Dorthea Collier, their teacher, earned twelve dollars per week and was responsible for twenty-four pupils for whom education was compulsory from ages seven through fourteen. Admission to high school was competitive.

A new one-room schoolhouse was built in 1903 at a cost of $2,684.73, but it was used for only about a decade before one-room schoolhouses in Falmouth were shut down and the pupils transported to other schools by very simple buses.

DEMOGRAPHIC AND OCCUPATIONAL CHANGES

In 1860, although farming was still the most common occupation, ten male Hatchville residents were employed as carpenters, shipbuilders and mariners. For the next several decades, Hatchville men, mostly from traditional farming families, would find work opportunities in the maritime trades. A mill in 1860 employed one female resident in "factory operations," and one male was a dyer. Still, some residents carried on the important trades of cooper, wheelwright, painter and mason.

In the census of 1880, there were still fourteen families in Hatchville, but only nine were the same as in 1860. Even though most of the established families were farming, the age of the heads of households was increasing, which was probably because elder mariners returned to live on family farms. Comparison through the years showed that within a family the males had different jobs at different times of their lives. For example, brothers Samuel and Charles Turner were mariners in 1870, but later, they became a wheelwright and cranberry grower, respectively. Edmund Chadwick Hatch (1816–1855) farmed and later sold stationery. He had two sons: the younger son, Emerson, was a mariner but later became a commission merchant, and the elder son, Ebenezer, began as a mariner but later, after marrying, did several jobs. Ebenezer's son, Allen B. Hatch, born in 1886, the last of the clan to reside in historic Hatchville, recalled that when he was a boy, they hunted red fox for their pelts to earn cash during the winter and he and his father peddled berries, garden vegetables, milk, eggs and chickens.[17] In 1919, Allen lived with his mother Adelia Clapp Hatch, then age seventy-three, on Hatchville Road but later married Marion Davis. They had no children. He was a farmer, carpenter and house painter.

Prior to the twentieth century, the Hatchville Turners and Robinsons began to lose members. The times were changing, and commerce was drawing men and women off the farm. Cape Cod had about one thousand manufacturing jobs in textiles, shoes and bricks around 1890, but very soon thereafter, these businesses couldn't compete with larger manufacturers.[18] Ultimately, the large related families began to fragment, stop or reduce their farming and sell some of their land to outsiders.

For many years, the Turners were a large family living along the southwest side of Coonamessett Pond on what was Meeting House Road (now Hatchville Road). Their clan was large because Japheth Turner (1793–1863) and his wife, Susanna Fish (1798–1878), produced fifteen children; all but three survived for some time. Japheth Turner was a farmer, but his sons, Charles H. Turner (1840–1914) and Zenas Turner (1819–1876), were both mariners. Zenas also farmed, but his sons, George and Harry, eventually gave up farming. In 1884, George Turner became a grocer and baker in Falmouth village, a job he held for twenty years.

In 1898, Harry Turner sold the Turner farm on Hatchville Road and moved to Falmouth village, where he became a constable. The new owners were John and Lura Randall of Wayside Farm, who raised and sold pigs. Samuel Turner (1837–1914) was a mariner, a veteran of the Civil War and the last Turner in Hatchville. He made and repaired wagon wheels on his farm on Crooked Pond on Main Street (later Sam Turner Road) near his elder brother, Zenas. After he died, his farm was sold to Reverend George Lawrence Parker.

The Robinsons, who were businessmen and farmers, were an important family in Falmouth. One of their own, Isaac Robinson, was one of the founders of Succanessett. The Hatchville Robinsons, who lived south of Coonamessett Pond adjacent to the river bogs and the East End Meetinghouse, had a similar family history to the Hatchville Turners. Ezekiel Robinson (1731–1805) and his wife, Hannah Gifford (1724–1801), also had fifteen children. Three of their sons, Charles, Seth and Bartlett, farmed too. However, Charles's son Peter (1780–1866) and Seth's son Freeman (1802–1879), who was married to Temperance Hatch, were both mariners. Nevertheless, Bartlett's son Nathaniel (born 1808) was still farming in 1880, as was his son Edwin, then twenty-seven, who continued for at least twenty more years. William F. Robinson (born 1842), son of Freeman and Temperance, may have been the last Robinson in Hatchville.

The Geggatt family is a link between the older and newer residents. They settled in Hatchville in the nineteenth century and continued to farm there into the twentieth century. John Geggatt (1805–1880) was a farmer who resided near Shallow Pond south of current Thomas Landers Road. He and his wife, Melory Ann Fish, had six children; the youngest, Ezra F., was born in 1852 and died in 1937. At seventeen years old, Ezra went to sea as a whaler and returned to Hatchville to farm with his elderly father. Ezra married eighteen-year-old Clara H. Fish in 1894, and they had four children. Their farm was known for its apples and especially for its cider, as well as for the sale of vinegar, small fruits and fertilizer.

CHANGING TIMES AND LAND OWNERSHIP

The industrial age spun up some great innovations, new education and training opportunities, modern agricultural methods and substantial wealth. In Falmouth, by 1910, the main business district had electric streetlights, and some villages, but not Hatchville, had updated water supplies. The town also had a new inner harbor and became known as a summer resort for the wealthy.

At this time, Hatchville's farming families faced a new reality: there were better opportunities elsewhere for the younger generation. Less expensive imported food was replacing locally grown food, which made farming less profitable. Additionally, the average age of residents was increasing. Hatchville still had relatively inexpensive land, and some of the inhabitants decided to sell for a quick profit; the turnover was large and amazingly fast.

They sold their land to several individuals from the Boston area and to a new business, the Coonamessett Ranch Company, whose directors were members of the family circle of Chicago industrialist Charles R. Crane.

Chapter 2

CRANE FAMILY CIRCLE

Beginning in 1915, Hatchville farmers sold their land in earnest. There were two kinds of sales—to unrelated individuals and to members of the family circle of Chicago industrialist and Woods Hole summer resident Charles R. Crane. The scale of the sale to individuals, although hundreds of acres, was dwarfed by the sale of thousands of acres of land and some farms in the northern end of Hatchville to members of the Crane family circle.

To understand the subsequent actions of the Crane family circle, it is important first to know why and when Charles Crane even took an interest in the picturesque seaside village of Woods Hole. That he did and that he later became a summer resident of Woods Hole is the first and most important improbable event in this narrative.

Charles R. Crane (1858–1939) was the eldest child of Richard Teller Crane (1832–1912), the founder of the highly profitable Chicago-based Crane Company. The Crane Company, which began as a simple metal foundry in 1855, within fifty years evolved into a multinational manufacturer of lightning rods, railroad equipment, pipes, valves, plumbing and bathroom fixtures. Charles worked in the family business and traveled frequently to Europe and Asia on business, but just as often, he was discovering new people and cultures there.[19]

In a departure from his usual travel routine, in 1900, Charles Crane visited Woods Hole, a fairly prosperous village of Falmouth. All the buildings then were wooden and the roads dirt. At the most southwestern tip of Cape Cod between Buzzards Bay and Vineyard Sound, Woods Hole was a two-hour

train ride from Boston and, then as now, a forty-five-minute ferry ride to the island of Martha's Vineyard. It was home to two scientific institutions, the U.S. Commission on Fisheries and the private, largely university-supported Marine Biological Laboratory (MBL), which was founded in 1888. The MBL is where scientists still come to explore life's processes by collecting and studying simple sea creatures, and it is where Crane's younger sister, Frances (an MD), and her husband, Frank Rattray Lillie (a PhD), a Canadian zoologist on the faculty of the University of Chicago, spent their summers.[20]

In order to help this young institution so important to the Lillies, Charles Crane joined the board of trustees in 1901 and served as president of the MBL Corporation from 1904 to 1924. As a trustee, Crane donated land and funded construction of the first brick building, which bears his name, and importantly, he donated $1.25 million for operations and endowment. By the end of his term as a trustee, funding for a new laboratory building, named for his brother-in-law, Frank Lillie, was assured.

Opposite: Charles R. Crane and an unidentified child. *Courtesy Woods Hole Historical Museum Archives.*

Left: Wilfrid Wheeler, Massachusetts agricultural administrator, Concord and Hatchville farmer and horticulturist (circa 1920). *Courtesy Massachusetts Horticultural Society.*

Wilfrid Wheeler was a member of the Crane family circle because he was related to the Lillies. His wife and first cousin, Canadian-born Emily Lillie Wheeler, was the sister of Frank Lillie. The two families were very close, and the proximity of Woods Hole to the Wheelers' home in Concord, Massachusetts, permitted the families to visit more frequently than an occasional long trip to or from Chicago.

Wilfrid Wheeler was ambitious and smart. He was aware that he had a lot to contribute to society and not just in agriculture. He had studied botany at Brown University and horticulture at the Bussey Institution of Harvard University and at the Royal Botanic Gardens at Kew in London. For about fifteen years, he farmed in Concord on the family homestead. Like his father, George, he perfected methods of growing and transporting to the Boston market high-profit crops such as strawberries, asparagus and cut flowers.

In recognition of Wheeler's agricultural skills and business acumen, he was chosen several times as "Farmer of the Year," which led to his appointment in 1913, at the age of thirty-seven, as a member of the Massachusetts Board of Agriculture. The next year, his colleagues elected him secretary of the board. Wheeler was popular with farmers. Here was one of their own charged with bringing both modern agricultural methods and farm family prosperity to a region of the country that had seen better days. Needless to say, Wheeler was intent on reinvigorating the farmland after decades of urbanization. He found a receptive audience in Charles Crane.

In 1910, Charles Crane bought Daniel Webster Butler's house on eight acres of Juniper Point (a small jut of land between the jade waters of Little Harbor and the azure of deep Great Harbor in Woods Hole) for $70,000. He then renovated the home, and it became the locus for his extended family circle, especially in the summer months. It also meant that Charles Crane and Wilfrid Wheeler would have many more opportunities to meet and share their ideas about progressive farming and the state of agriculture. One such opportunity was when they planned for Wheeler to landscape Crane's Juniper Point compound in 1912.

Woods Hole was once a coastal trading and whaling center and briefly, until 1889, the location of a guano factory. Part of this small village soon became a seaside resort with expensive stately summer homes for bankers and corporate executives laid out on a private road on the former fertilizer factory site on a snake-like peninsula called Penzance. Equally impressive was the "airplane house" at the tip of Juniper Point, which the generous father Charles Crane commissioned in 1912 for his daughter Mary Josephine Bradley and her family.

The Crane family circle's documented involvement in the future of the village of Hatchville began in 1914. Two years after the death of Crane Company founder, Richard Teller Crane, Charles sold his share in the Crane Company to his brother Richard for about $14 million (over $200 million in current dollars) and retired from business at age fifty-six. While not directly involved in most of the other actions that followed, Charles Crane, who paid for most of them, was kept informed as he flitted to and from all points on the globe from his Fifth Avenue office in New York City, where his trusted staff carried out his wishes.

Charles Crane's retirement as president of the Crane Company directly affected two members of his family circle who were employed there and tied to him, and thus needed to find new careers. One of them was Crane's son-in-law, Robert W. Leatherbee, who was born in West Newton, Massachusetts, in 1881 and, in 1905, graduated from Harvard College. According to a family genealogist, the Leatherbees were descendants of

Aerial view of Juniper Point, Woods Hole. The "airplane house" is at the tip, the Crane compound is in the middle and the Coast Guard Station is to the right, on Little Harbor. *Falmouth GIS.*

Leatherbee family, circa 1916, with Robert and Frances Anita Leatherbee and their three sons on the left. Frederick Leatherbee is standing on the far right. *Courtesy Harriet Needham.*

French Huguenots called Larrabee who settled near Boston. In 1915, Robert's farther, Charles W. Leatherbee, a Boston lumber merchant, was head of the family. Charles and Harriet Felton had four children: Robert W., Frederick K., Clifton and Eleanor. Charles Leatherbee was a founder of the Brae Burn Country Club in West Newton, and Robert Leatherbee named his farms and yacht *Brae Burn*.

In 1907, Robert married Chicago-born Frances Anita Crane, the third child of Charles R. and Cornelia Crane. While in Chicago, the Leatherbees lived at the Crane's stylish multistory Michigan Avenue home. At that time, Robert Leatherbee worked at the Crane Company, first as a superintendent of the Malleable Iron Department and later as the general superintendent of manufacturing.

Leatherbee continued at the Crane Company until 1914, when he and Frances Anita and their three sons (Charles, Richard and Robert) relocated to Woods Hole. This move brought Robert Leatherbee closer to his younger brother Frederick K. Leatherbee, a Harvard-educated investment counselor, and the other family members. Both Robert and Frances Anita, two big personalities, would play significant but not overlapping roles that defined land use in Hatchville.

The other Crane Company employee seeking a new career was Charles R. Crane's elder son, Richard Teller Crane III, who was born in Denver, Colorado, in 1882, graduated in 1904 from Harvard College and from 1910 to 1914 served as president of the Crane Valve Company in Bridgeport, Connecticut. In 1909 he married Ellen Douglas Bruce of Berry Hill, Virginia, and lived in Chicago. About the time that his father retired, Richard ran unsuccessfully as a Democrat for a seat in Congress for Illinois. His next appointment, in 1915, was as secretary to U.S. secretary of state Robert Lansing in the Wilson administration. While this may seem rather unlikely, it actually was quite an easy move: Charles Crane had been the major financial backer of Woodrow Wilson's presidential campaign in 1912. Wilson even considered Charles Crane as ambassador to Russia and as a member of the newly created Federal Reserve Bank. Charles Crane declined both offices in order to concentrate on the Crane Company business after his father's death. Richard would have a brief role in the events in Hatchville.

Although not in need of employment, the youngest of the four Crane children—John Oliver Crane, who was born in 1899—was at the Asheville School in North Carolina when his father retired. Not until the 1930s did he play an active role in the events in Hatchville.

In summary, in 1914, Robert and Frances Anita Leatherbee were living in Woods Hole, Charles and Cornelia Crane in New York City and Wilfrid and Emily Wheeler in Concord.

Chapter 3
COONAMESSETT RANCH COMPANY

Charles Crane and Wilfrid Wheeler met on December 19, 1914. As reported in the *Falmouth Enterprise*, the Honorable Wilfrid Wheeler of the state board of agriculture was on a business trip to Charles Crane's home on Juniper Point in Woods Hole. There's no record of what they talked about when they likely sat before a warming fire. It would have been uncharacteristic of the well-traveled and knowledgeable Crane if he had not commented on the war in Europe. He had been traveling throughout Europe since he was nineteen years old, and beginning in 1887, he traveled to the Russian Empire on Crane Company business, after which he became a prominent Russophile.

One topic of conversation between the bespectacled Wheeler and the tall, goateed Crane that day certainly was the 1,487 acres of unused lots in the towns of Sandwich and Bourne that Wheeler had purchased at about five dollars per acre with Crane's money. He bought this land from the Hannahs' "Homestead Trust," a real estate marketing company. There must have been an understanding between Crane and Wheeler that purchasing these parcels (and eventually more land) could be both a business opportunity and a societal benefit. Both were conscious of their obligation to do good.

Opposite, top: Crane house on Juniper Point, Woods Hole, the site of family gatherings. *Courtesy Woods Hole Historical Museum Archives.*

Opposite, bottom: Sandwich, Massachusetts sheep pasture lots and woodlots purchased by Wilfrid Wheeler from the Homestead Trust. *Courtesy Sandwich Engineering Department.*

FALMOUTH
SNAKE POND
MASHPEE

These Sandwich and Bourne lands that Wheeler had purchased, and would add to, were two-hundred-year-old deeded twenty-acre woodlots and large sheep pasture plots. The woodlot owners had sold cordwood for home use and, later in the nineteenth century, sold wood to the Sandwich glass industry to fuel its furnaces. When the Sandwich proprietors divided these twenty thousand acres of common land among the townsmen, Sandwich's western boundary was Buzzards Bay. It wasn't until 1884 that the town of Bourne was created from land to the west of the north–south rutted Turpentine Road, which was a stagecoach route for postal delivery to Falmouth at least as early as 1823 and surely ended at the Hatchville post office after 1858. The road was named for the distilled pine resin harvested nearby.

Now we come to the second improbable event. Following Wheeler's initial land purchases, the Crane family circle organized and bought thousands of acres of contiguous land in Falmouth, Bourne, Sandwich and Mashpee, in which would fit the island of Manhattan or the Cape Cod town of Provincetown, Massachusetts. All this land would be owned by a new Massachusetts corporation—the Coonamessett Ranch Company, backed financially by Charles R. Crane and directed by this family circle.

The Crane family circle gathered at Crane's home on Juniper Point in April 1915. By the summer, Frederick and his wife, Dorothy Leatherbee, had moved to North Falmouth. In the next months, Fred Leatherbee would buy land and farms, including associated tillage and pasturage north and east of Coonamessett Pond. While a magnificent home on three acres overlooking Buzzards Bay called "White Crest" was being built for them by Charles Crane, Robert and his wife, Frances Anita, rented a home in Woods Hole. Robert, who had owned a four-hundred-acre gentlemen's farm in Lake Forest, Illinois, was advising his brother Fred.

Charles Crane likely favored the agricultural aspect of the venture for two reasons: first, because he saw it as a way to help Frances Anita and Robert Leatherbee make a new life at their adopted home in Falmouth and, secondly, because Crane had a keen interest in agriculture. Locally, he had a ten-acre productive farm, which he had begun in 1915 along both sides of Woods Hole Road about a mile north of his summer home, and he also developed an extensive date farm in the Coachella Valley of Southern California.

Acquisition of the Sandwich and Bourne lands may have been attractive to the family circle because of the recent opening of the one-way, privately owned Cape Cod Canal. The canal permitted ships headed to Boston from

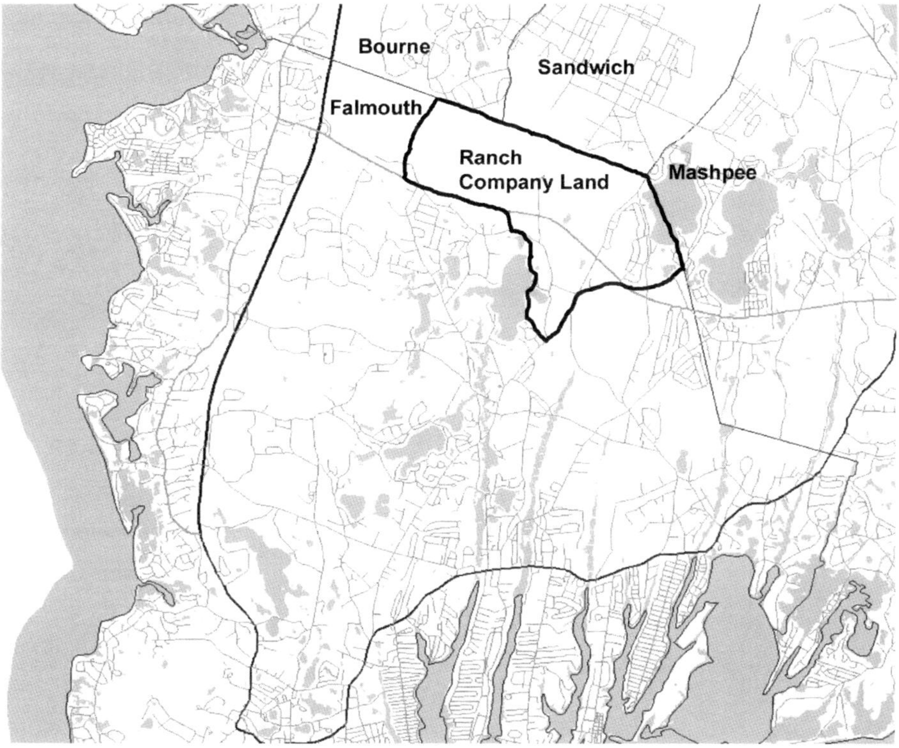

Coonamessett Ranch Company's 2,500 acres in Hatchville in 1924. *Produced by the author from a map courtesy of Tom Stone.*

ports on the East Coast to reduce the time for the journey while also avoiding the dangerous sandbars off eastern Cape Cod.[21] Optimistic predictions that the canal would bring population growth and more manufacturing to Sandwich never materialized. The unprofitable canal was purchased by the federal government and turned over to the Army Corps of Engineers in 1928.

WHEELER'S PLAN EXPLAINED

As Wilfrid Wheeler told the *Boston Globe* in 1918, he had a plan for the use of this land: he envisioned a colonization scheme to increase the number of farms and farmers of Cape Cod. The prospective farmers would learn the principles of scientific farming at a model farm in Hatchville. After training, the new farmers would become eligible to purchase tracts of cleared woodland that had been planted and improved from the Coonamessett Ranch Company on long-term payment plans. Wheeler, a practical visionary, also saw this model farm, surrounded by nearby smaller farms, as a solution to the continuing loss of farms in Massachusetts.[22] Hatchville was a desirable location for the model farm because it was a settled farming community with sufficient tillable flat land and deep wells.

For many years, the purpose of the Coonamessett Ranch Company was mysterious and led to public confusion. This may have been due to the headline for the article in which Wheeler explained the model farm, which boasted: "Cape Cod Has Biggest Farm East of the Mississippi." The model farm, at less than five hundred acres, was much smaller than the eleven thousand acres the Company had accumulated by the time of publication.

ACTUAL LAND PURCHASES

With Charles Crane's money to spend freely, between June 1915 and September 1917, Frederick Leatherbee bought sixty parcels of land totaling 2,936 acres, mainly in the northern part of Hatchville. Wheeler transferred his aggregated purchases of 8,702.68 acres in Sandwich, Bourne and Mashpee to Frederick Leatherbee on September 28, 1917. Within a week, Frederick Leatherbee transferred 11,638 acres to the recently formed Coonamessett Ranch Company.[23]

During 1918, Frederick Leatherbee purchased nine separate parcels totaling 269 acres in Falmouth, Sandwich and Bourne, which he also transferred to the Coonamessett Ranch Company on August 14, 1920. By this time, he had moved to Port Chester, New York, where he served as vice-president and treasurer of Mallory Industries, a boiler manufacturer. Additional land purchases in the next year, including adjacent cranberry bogs, brought 14,000 acres (twenty-two square miles) of contiguous land under the control of the Coonamessett Ranch Company.

The settled and improved Hatchville land cost considerably more than the woodlots purchased by Wheeler in Sandwich and Bourne. For example, a farm, owned by Will Chadwick, overlooking Coonamessett Pond cost $7,418 and a one-hundred-acre farm to the east of the Pond owed by the Manuel Ignos cost $4,236. Pastureland and cranberry bogs were more expensive too.

Who were the people who sold their land and farms to the Coonamessett Ranch Company? The Hannahs, from whom Wheeler bought nearly half of the company's land holdings, were definitely land speculators. Some of the early Hatchville settlers who sold some of their land included the Robinsons, Hatches, Fishes, Fullers and Fishers. Some of the recent immigrants from the Portuguese Azores and Cape Verde, such as the Dutras, Damarals and Sozas, also sold their land.[24] One can only infer that these people sold because the time and price were right.

COONAMESSETT RANCH COMPANY

The Coonamessett Ranch Company was incorporated in the commonwealth of Massachusetts on September 8, 1916.[25] At the time of incorporation, the company directors, who were very minor investors and probably surrogates, were Norman J. MacGaffin (West Medford), Albert A. Richards (Boston) and E.M. Churchill (Winthrop Highlands). Their business address at the time of incorporation was 53 State Street, the Boston Stock Exchange, but the company's place of business was listed as North Falmouth (Frederick Leatherbee's temporary home) with an alternate site in New York City (Charles Crane's home). None of the three incorporators ever owned land on Cape Cod. From the foregoing, we can infer that they were acting as agents of Frederick Leatherbee, who had been an investment counselor and had worked at State Street brokerage firms prior to moving to North Falmouth.

The stated purposes of the Coonamessett Ranch Company were quite broad and included agriculture and timber interests, as well as land sales and development, none of which was inconsistent with Wheeler's model farm plan. The company was valued at $200,000 by the surrogate directors, who issued $150,000 in preferred and $50,000 in common stock shares, each at a par value of $100. The trio subscribed to only ten shares of common stock. There are no records in Massachusetts to show when these surrogates resigned as directors.

SIMULTANEOUS HATCHVILLE LAND PURCHASES

While the Coonamessett Ranch Company was buying land and farms on the north and east part of Hatchville, farms and land were also being purchased on the south and west of the small village. Initially, these unrelated buyers had one thing in common: their interest in birds, namely, duck hunting and chicken farming.

In addition to its agrarian milieu, Hatchville had acquired a deserved reputation for excellent freshwater fishing as well as duck and deer hunting, especially from sportsmen from around Boston. Most touted was fishing the Coonamessett River for sea-run brook trout, a once populous fish along the Atlantic coast. Back before the streams were dammed and the water temperature rose, this fish would spawn in the cool waters and return to the sea for a few months in the summer. Sea trout were plentiful enough to be sold door to door in Hatchville by local Native Americans.

Waquoit innkeeper Asa Phinney Tobey ran the Huntsmen's Paradise, a destination for discerning sportsmen. Apparently, in order to enhance the fishing experience in 1884, the tenth anniversary of his lodge, he obtained ten thousand salmon fry for stocking Coonamessett Pond. No salmon were found in July 1948.

George St. Amant, a Boston-based wool merchant and importer of Egyptian cotton who resided in Newton, Massachusetts, was a duck hunter. In 1915, he bought sixty-six acres on the southwest side of Coonamessett Pond from the estate of Joseph Story Fay to establish a seasonal duck hunting preserve. To support his hobby, St. Amant built a lodge and a shooting shelter along the pond, called a camp. But he had more ambitious plans, and during the 1920s, he expanded his land holdings, harkening back to elite Bostonian gentlemen's farms of the previous century.[26] St. Amant removed pine trees destroyed by gypsy moths on the northwest part of his property to create grazing meadows for prized Guernsey cattle. In 1920, he paid $10,000 for the bull "Milk Man Ultra King"; though not a record, it was expensive. In 1926, St. Amant bought an additional forty acres along Hatchville Road from Lucina Swett and continued to buy costly Guernsey cattle for his dairy farm. Eventually, his Atamannsit Farm boasted eighty cows, forty young and a couple bulls, all housed in three modern barns. The property also had an engine house and an icehouse. But in 1929, St. Amant, who was apparently in financial difficulty, was forced to sell his fine herd at auction.

Urban-based hunters and fishermen and the Coonamessett Ranch Company were not the only people interested in Hatchville land; some folks

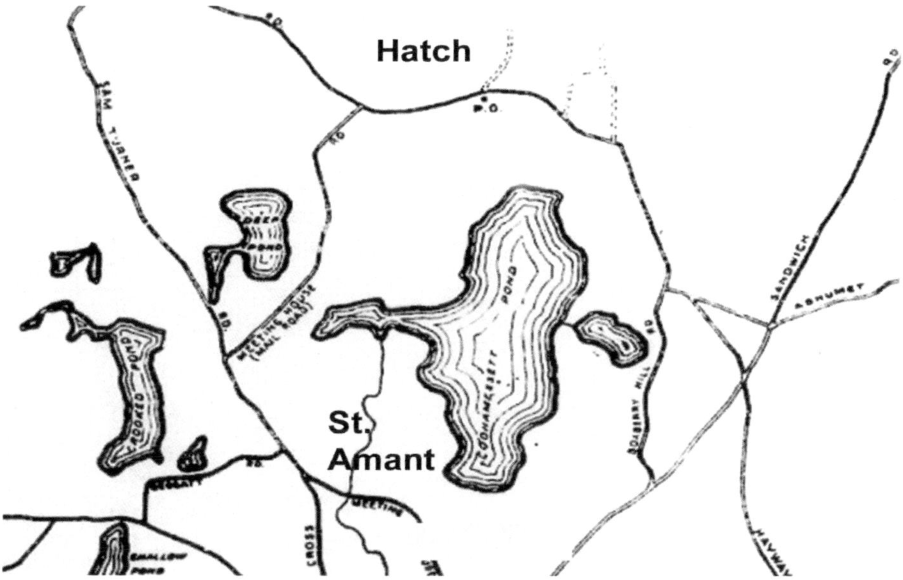

Land on the southwest side of Coonamessett Pond purchased by George St. Amant in 1915. *Annotated by author from a map in public domain.*

left the cities to experience the rural life and landscape. One of the first was George Edward Tent, who was born in 1874 in Chelsea, Massachusetts, and studied business at Dartmouth College, graduating in 1897. In 1916, George and his wife, Marion, were looking for a change from life in suburban Boston and his job as a cashier at New York News Company. They wanted to start a poultry farm. According to Tent's granddaughter, the late Margaret Hopkins Freeman, "they had vacationed in North Falmouth and thus knew the town and real estate opportunities."[27]

Model Farm (1917–1919)

The Coonamessett Ranch Company's valuation was accurate. In historic Hatchville, a very expensive modern dairy and truck farm rose slowly on the flat land purchased from the Hatch family and their neighbors. No longer were oxen employed to clear the land; now two leased, ten-ton Holt caterpillar-tread tractors called "peace tanks" both cleared the vegetation and prepared the soil. Still, Wheeler and the Leatherbees budgeted $1,000

Rear of Will Chadwick's farmhouse at the north end of Coonamessett Pond. The farmhouse was renovated for the Wheelers; later, it was a teahouse and then the Coonamessett Inn. *Courtesy Harry Donald.*

per month for the labor of sixteen men to clear an additional six hundred acres of land. Meanwhile, the spacious Chadwick farmhouse, which overlooked the north end of Coonamessett Pond and the bog, was renovated for the farm manager's residence.

Through June 6, 1917, according to a memorandum between Richard Crane and his father, $201,000 (about $2.5 million today) of Charles Crane's money had already been invested in the Coonamessett Ranch Company and an additional $13,250 and $3,750 paid to Robert W. Leatherbee and Wilfrid Wheeler, respectively.[28]

Between July 1, 1917, and June 30, 1918, Fred Leatherbee and Wheeler, who was still secretary of the board of agriculture, spent $4,000 per month to make the farm truly functional. Fortunately, the new farm had a good water supply, including a large, deep well pumped by a windmill. They anticipated selling milk, butter, hens, hogs and even cultivated cranberries from their bog. Nevertheless, the farm still required a subsidy of $25,000 from Charles Crane. A fair estimate of Crane's investment in the Coonamessett Ranch

Company's farm and land purchases would be $275,000, or at least $4 million in current dollars.

"[Robert Leatherbee's] mind is full of cows and stumps," U.S. secretary of agriculture David F. Houston, who summered on Juniper Point, wrote to Charles Crane in August 1917.[29] Robert would soon take time away from implementing the farm plan in Hatchville to serve in Washington, D.C., with the War Shipping Board, the federal government agency that built noncombatant ships.

On January 14, 1919, four days before the start of the Paris Peace Conference ending World War I, the real directors (the family circle) met at Crane's New York City apartment at 31 West Twelfth Street and issued shares of the Coonamessett Ranch Company stock. As befitting the major investor, Charles Crane received most of the preferred stock. Charles R. Crane, Robert W. Leatherbee, Frederick K. Leatherbee and Wilfrid Wheeler, who was the secretary/treasurer, received common stock.[30] At this meeting, Wilfrid Wheeler accepted a five-year contract to become the farm manager for the Coonamessett Ranch Company. The contract changed the direction of Wheeler's career. By accepting this position, he was poised to implement his farming plan on a grand scale with methods new to Cape Cod.

After the director's meeting, Charles Crane attended the Paris Peace Conference, where he was drafted by President Wilson to head the King-Crane Commission to assess the postwar wishes of the Arab populations in the Middle East. For the next twenty years, Charles Crane was the most vocal Western advocate for Arab self-determination, especially in Palestine.

The founding of the Coonamessett Ranch Company is no longer opaque. We know now who started the company, who bought and paid for the land and the model farm and about Wilfrid Wheeler's central role.

Chapter 4

WILFRID WHEELER

MODEL FARM AND ASHUMET FARM

After the death of Silas Hatch in 1919, Wilfrid Wheeler and Robert Leatherbee, both directors of the Coonamessett Ranch Company, became two of the most influential men in Hatchville and later in Falmouth. For the next five years, they still were involved with the Coonamessett Ranch Company, then each became successful in agriculture in their own way—Wheeler as a horticulturist who developed, grew and sold specialty crops and later American holly, Leatherbee as a dairy farmer producing quality milk.

While he was secretary of the board of agriculture from 1913 to 1918, Wheeler literally had to watch over the birds, bees and cows (through the Dairy Bureau). The stated purpose of the board of agriculture, which was formed in 1852, a decade before the USDA, "was to represent fairly, every class of agricultural knowledge in the state" and "to promote crop and animal husbandry."[31] The secretary of the board managed the department and was the face of agriculture to the public and to his peers in Massachusetts.

In this job, Wheeler promoted the growing and marketing of small fruits (strawberry, gooseberry, raspberry, blackberry and currant) and particularly touted the success of the strawberry business in Falmouth for the last two decades. In 1916, Wheeler and the extension agent in Barnstable County helped the mainly Portuguese strawberry growers in Falmouth establish the Cape Cod Strawberry Association. By organizing, these farmers were able to better market and transport their valuable fruit to urban buyers. In Falmouth, strawberries were grown on hundreds of small plots and cared for and harvested by family members. Lack of financial credit, which was an important issue for these strawberry farmers, was made available through the intervention of local landscape gardener and florist H.V. Lawrence as well as two new fellows, Frederick K. Leatherbee and Robert W. Leatherbee.

Wheeler, as board secretary, also tackled the issue of milk marketing in Massachusetts. Everyone wanted safe, disease-free milk. Analysis showed that most of the milk consumed or processed in urban areas like Boston, even though shipped by rail, was less expensive than locally produced milk. The reason was obvious: dairy farms had been displaced by farms producing the more profitable fruit and vegetables for the urban market. Centralization of the dairy industry, Wheeler noted, was rampant, which contributed to the decline of local dairies. However, he saw one bright spot: "Where increases are noted, this is due to the effect of local cooperative creameries or to an increase in price guaranteed by the dealers."[32] Knowledge of the milk industry would serve him well on the model farm in historic Hatchville.

WILFRID WHEELER FIRST COMMISSIONER, DEPARTMENT OF AGRICULTURE

Governor Samuel Walker McCall, a former Republican congressman, had appointed Wilfrid Wheeler as one of fourteen members of the newly formed Department of Agriculture, effective September 1, 1918. At their first meeting on September 4, 1918, the other members, most of whom were farmers, elected Wheeler as their commissioner for a three-year term. By statute, he was paid a salary of $3,000 per year. Wheeler's term as commissioner lasted one year. A massive reorganization and consolidation authorized by an amendment to Massachusetts's constitution and adopted by referendum in 1918 led to six department heads being sacked by governor-elect Calvin Coolidge in November 1919.[33]

Wheeler likely knew in January 1919 that he would not be able to keep his job as commissioner, and so when given the opportunity, he chose to personally implement his plan to increase the number of farms and farmers on Cape Cod as manager of the model farm.

Boston Daily Globe, November 25, 1919

GOVERNOR NAMES NEW STATE BOARDS
MANY WELL-KNOWN MEN LOSE PLACES

Cole Heads Public Works
--Gen Sweetser, Labor
And Industries

SOME OF THE GOVERNOR'S APPOINTEES
TO THE MORE IMPORTANT COMMISSIONS

Gilbert For Agriculture Chief;
Attwill, Public Utilities;
Foote, Public Safety

Among Those Dropped Are Rane,
Wheeler, Sohier, McNary,
Rothwell, Macleod

Boston Daily Globe article describing turnover of heads of commissions and boards. Wilfrid Wheeler was replaced as commissioner of the Department of Agriculture. *From the* Boston Daily Globe, *November 25, 1919.*

COONAMESSETT RANCH COMPANY ACTIONS, 1919

On April 28, 1919, the Coonamessett Ranch Company directors elected Robert W. Leatherbee president, while his brother Frederick resigned as a director and left Falmouth. Charles R. Crane received additional stock for the $35,800 he had paid into the company through April 1, 1919, when the company assets were valued at $331,220, mostly as land, livestock, buildings and equipment.

After the April directors' meeting, Frederick Leatherbee, no longer a director, inquired about selling his shares of common stock within the company. Richard Crane thought it was a bad idea; the company was "no where near being on a self-supporting basis, and its success largely depends on future developments. I would also as a Director not favor the purchase of the stock by the Corporation."[34]

Richard Crane resigned as a Ranch Company Director on May 19, 1919, to become the first U.S. ambassador to Czechoslovakia. He took no further direct interest in the company. He died on October 3, 1938, at the age of fifty-six from a fatal gunshot to the right temple at Westover, his Virginia estate, while hunting sora.[35]

MODEL FARM MANAGER, COONAMESSETT RANCH COMPANY (1919–1924)

For the remainder of 1919, Wheeler spent weekends at the farm and returned to Boston during the week to carry out his duties as commissioner and look in on his Concord home and farm. During the warmer months, Wilfrid, Emily and their two younger sons—Charles Lincoln, five, and Richard Parkman, two—lived in the renovated Chadwick farmhouse but moved to a house in Woods Hole during cold weather. The boys were home-schooled by their mother until they reached high school age. While Emily handled their education, Wilfrid taught them farming, how to build and care for things and how to shoot and handle guns.

Under Wheeler's management, the model farm operations grew. Machinery was introduced to feed corn into the two-hundred-ton silos as well as to shell and sort peas. He hired a new foreman, C.F. Hubbard, and paid him $1,500 a year plus housing. A new greenhouse for starting tomatoes and cucumbers was built and the canning department enlarged to accommodate the production of fruits, vegetables, jams and jellies and honey.[36]

Milk from Holstein cows, which was produced at a rate of eight hundred quarts per day and sold for fifteen cents per quart, was delivered to the shore

WILFRID WHEELER, *Treasurer and Manager*
CLIFFORD L. HUBBARD, *Assistant Manager*

COONAMESSETT RANCH COMPANY

Producers of Highest Quality
MILK AND CREAM
EGGS AND POULTRY
FRUIT AND VEGETABLES

Express and Freight: North Falmouth, Mass.
Telephone: Falmouth 256-2

A Part of the Herd

Breeders of
HOLSTEIN CATTLE
CHEVIOT SHEEP
WHITE AND BARRED ROCKS

Post Office
HATCHVILLE, MASS.

Coonamessett Ranch Company letterhead used between 1919 and 1924. *Courtesy of the Falmouth Historical Society, Falmouth, Massachusetts.*

A 1922 advertisement of the Coonamessett Ranch Company. *From* Cape Cod and All the Pilgrim Land.

communities of North Falmouth, West Falmouth, Quissett, Woods Hole, Falmouth Heights, Cataumet, Pocasset and Gray Gables. The new paved road between North Falmouth and East Falmouth made delivery easier. The farm's retail stores in North Falmouth and in Falmouth sold eggs, broilers and roasting fowl and took orders for deliveries. Over 250 acres of diverse crops were planted, including rye, corn (60 acres), asparagus, oats, peas, grapes, buckwheat, millet, strawberry (5 acres), raspberry and potatoes, as well as a hay field.

Wheeler demonstrated that by the application of modern agricultural methods and piles of cash, diverse crops could be grown successfully in good soil on Cape Cod.

After moving to Hatchville, Wilfrid Wheeler quickly took leadership roles in several agricultural and horticultural associations on Cape

Cod. Information gleaned from the *Falmouth Enterprise* newspaper for 1920 shows the extent of his influence. Wheeler was chairman of the Barnstable Fair (August 31 to September 2), working with the Barnstable County Agricultural Society. In November, he organized the Falmouth Agricultural and Horticultural Society and became its first president (H.V. Lawrence was vice-president). In December, he ran a sectional meeting of the Cape Cod Farm Bureau and attended the Annual meeting of the Barnstable County Agricultural Society, where he served on the executive committee.

By 1921, the Coonamessett Ranch Company farm had twelve houses, twenty-five farm buildings, three water sources, two greenhouses and an irrigation system covering twenty acres. Despite the apparent postwar labor shortage, one hundred people were on the payroll. Wheeler improvised by employing young female students from agricultural colleges, who were called "Farmettes," in the canning plant, greenhouse and dairy.[37]

Nevertheless, the farm was still not self-sustaining, and in 1923, Charles Crane paid in $3,164.40 to support operating costs. The farm continued to function in this way until the end of the year, when Wheeler's contract expired and Crane's advisors counseled against continuing his financial support for the farm. The emphasis of the company's business shifted from farming to an inland summer resort centered on Coonamessett Pond, even though the farm still supplied produce, dairy and poultry to the company's stores.

WHEELER'S ASHUMET FARM AND FALMOUTH LIFE

According to Charles Wheeler, the shift from farming to a resort was a contributing factor to his father resigning as farm manager. Wilfrid Wheeler was a farmer and horticulturist, not an innkeeper. So in the fall of 1924, as his five-year contract as farm manager and his position as Coonamessett Ranch Company director were ending, Wheeler bought 320 acres and two buildings on an old farmstead south of Ashumet Pond on the county road (Ashumet Road) between Falmouth and Mashpee from the Coonamessett Ranch Company for $8,000.[38] The property had a coastal plain pond that would be a source of fresh water for nearby crop irrigation, above which was a south-facing slope. The house, which needed repair, had no conveniences. Electrical lines had to be strung. The site was treeless and the farm's fields overgrown. Eventually, tree stumps were removed by blowing them skyward

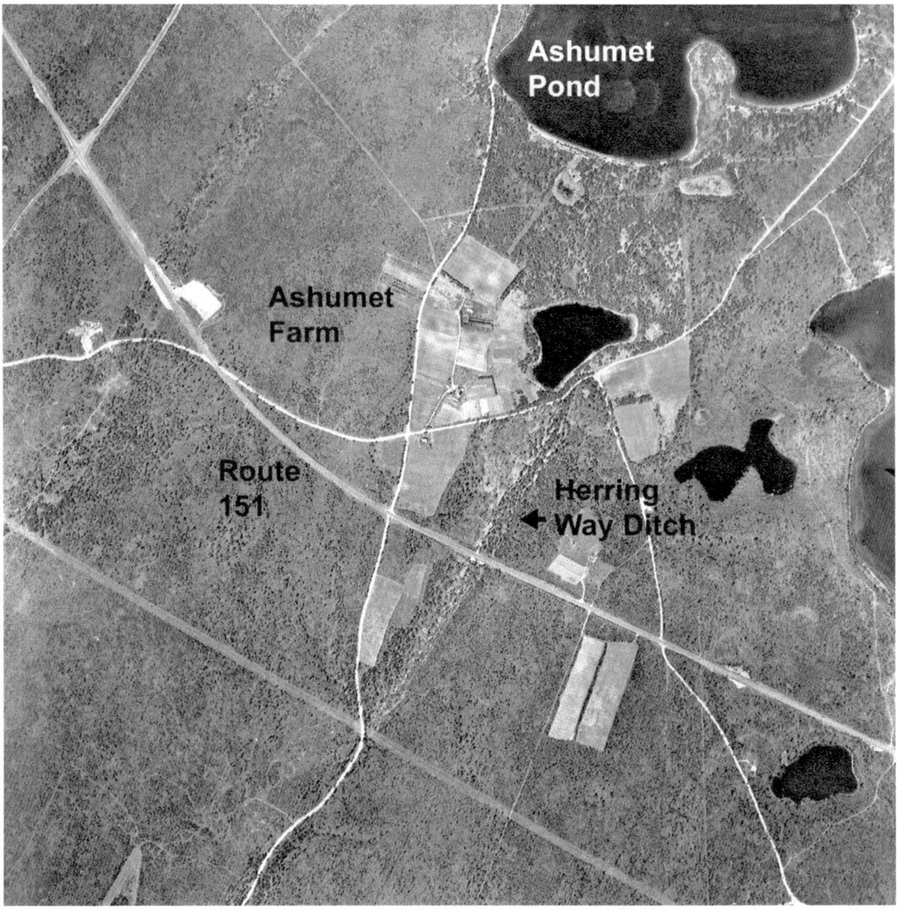

A 1950 aerial view of Ashumet Farm with coastal plain pond between the cultivated plots. The 1863 Herring Way ditch crosses under Ashumet Road. *Annotated by author from aerial photograph from Falmouth GIS.*

fifty feet with a dynamite charge beneath them, according to Charles Wheeler.[39]

It took much work to make the new farm habitable for the boys and their sophisticated mom, who was proficient in Greek, Latin, French and German. Known as a humanist, Emily Wheeler listened to classical music on a radio that her brother Charles Lillie had purchased for her. Wilfrid, an optimist as most farmers are, however, could be found during the baseball season listening to the Boston Red Sox play-by-play broadcasts. In contrast to Emily, Wilfrid was reserved and quiet. He was also a good listener.

By April 1925, Wheeler began to develop "Ashumet Farm," a truck farm, on part of the land near Ashumet Road. Charles Wheeler maintained that his father was then "fulfilling his horticultural destiny." Wilfrid hired Joseph Dias, who had worked for him on the Coonamessett Ranch Company farm, to supervise and prepare the new farm. Dias was really Wheeler's partner, and he continued to maintain the property for forty-seven years. They used a Fordson crank-start tractor in those days. And they had a truck to take the produce—at first strawberries, turnips, asparagus and sweet corn—to market. The farm had a horse and two cows, which were housed in a new barn. Meanwhile, Wheeler established home moving (a winter job) and construction, real estate and landscaping businesses to serve the large shore estates.

Even before starting Ashumet Farm and for decades later, Wheeler measured, recorded and reported to the *Falmouth Enterprise* newspaper the daily temperature and precipitation in Hatchville.[40]

Wheeler soon joined the Falmouth Board of Trade (a businessmen's association) and served as vice-president in 1928 and president in 1930, the same year he was president of the Cape Cod Horticultural Society. He also served on the Town Forest Committee.

Emily Wheeler was an avid gardener, educator and poet. During the 1930s, she served two terms as president of the Falmouth Garden Club and hosted meetings at Ashumet Farm. She also ran its junior gardener program in the public schools. In June 1936, after son Charles graduated from Harvard, Emily and her sister-in-law toured England and France for six weeks.

A congenial hostess, Emily often served a four o'clock tea for friends and family and was visited by various artists, among whom were Dorothy Norman, Georgia O'Keeffe and Alfred Stieglitz, but probably not all three at the same time. There was no alcohol at the Wheelers'; instead, apple cider was made on the farm. Wilfrid didn't smoke, but "fancy cigarettes" were available at Emily's soirées, according to nephew David Lillie. Dorothy Norman came to Cape Cod in 1937 and photographed various local landmarks, including the First Congregational Church in Falmouth village and the East End Meetinghouse in historic Hatchville.

Holiday gatherings, such as the one on Thanksgiving Day, brought the extended family to Ashumet Farm. Most of these feasts were prepared by a marvelous cook, Annie Howell, who was the Wheelers' maid for many years. While Emily may have supervised the dinner, Annie did the work. Annie and her daughter, Hazel, lived in small rooms near the farmhouse kitchen.

The Falmouth Garden Club visits charter member Emily Lillie Wheeler at Ashumet Farm in June 1931. *Courtesy of the Falmouth Historical Society, Falmouth, Massachusetts.*

As a farmer, it wasn't long before Wheeler advertised as a supplier of "fresh dug" strawberries, specifically "Howard No. 17," the predominant variety grown on Cape Cod, and he soon sold his cut flowers, lettuce and prize-winning melons in local produce stores. Wheeler had a kitchen garden as well as commercial plots. He also grew prize-winning specialty plants that he entered in many competitions and fairs. For example, in the 1930s, he entered Chesapeake strawberries, hot Hungarian peppers, miniature sunflowers, begonias, watermelon, lilies and tomatoes. In the next decade, he focused on lettuce, golden acre cabbage, Kendall melons, strawberry corn, *Montbretias*, Japanese onions and herbs from Emily's garden. Wheeler was anything but a conventional Cape Cod farmer bound by custom or convention. He mastered horticultural techniques to raise, select and even create, through cross-pollination, popular specialty crops.

Hollies and Beach Plums at Ashumet Farm

In 1930, a friend sent Wilfrid a dozen small American holly (*Ilex opaca*), which he planted on his property. These lovely plants grow in deep, fertile acidic soils to medium height in the understory but taller when exposed to sunlight. Thereafter, hollies became his major interest. Already concerned that vandals and harvesting of the boughs of berry-bearing female hollies for Christmas decoration had reduced holly numbers, Wheeler strived to protect the remaining wild plants and encourage holly plantings in public spaces on Cape Cod. Noting the rapid growth and hardiness of the plantings at Ashumet Farm, he embarked on a plan to commercialize the American holly.[41] Wheeler overcame the barriers to propagating and growing holly in New England by selecting winter-hardy plants with large fruit and pleasing form.

At the same time as he was perfecting holly propagation, Wilfrid Wheeler was also championing another iconic native plant, the beach plum (*Prunus maritima*). He collected, selected and propagated plants on Ashumet Farm for their fruit quality and their beauty. As part of his landscaping business, he planted these soil-holding plants along shore properties, especially after damaging storms. During the 1950s, Wilfrid Wheeler served on the executive committee of the Cape Cod Beach Plum Growers Association, an organization that tried to improve and commercialize the beach plum (used mainly in jelly) so that its popularity would rival that of the blueberry and cranberry.[42]

In the 1940s, Wheeler devoted much time and effort to the holly business. Locally, he planted English hollies at the Whitney Estate and majestic American hollies for Frances Anita Crane's home, White Crest, both in Woods Hole. After World War II, three successive displaced-person families handled the Ashumet Farm chores, which gave Wilfrid and Joe Dias more time to work with the hollies. One of these displaced European families who lived in a cottage on the farm was Latvian Alexander Cirulis and his wife and daughter. By the mid-1950s, Wheeler and Dias had transplanted and propagated one thousand stately American hollies and had named twenty-one varieties, including one called "Emily." His work on behalf of the native holly led to honors by the American Holly Society, the Massachusetts Horticultural Society and the University of Massachusetts College of Agriculture. Not content with just planting American hollies on the farm, Wheeler and Dias also planted Asian and European hollies, the rare *Franklinia* and the Japanese umbrella pine.[43]

Despite his demanding horticultural work, Wheeler continued his service to the town of Falmouth. The town's trees had been damaged by hurricanes and neglected during the war years. Wilfrid, the planner, saw the need for attention to this important part of the landscape of which Falmouth

Left: Wilfrid Wheeler's "Price List and Directions for Growing American Holly (*Ilex opaca*) in the North." *Coutesy Wheeler archives.*

Opposite, top: Elderly Wilfrid Wheeler examines one of his hollies in the winter. *Courtesy Wheeler archives.*

Opposite, bottom: Map of the Ashumet Holly and Wildlife Sanctuary showing locations of the memorial grove honoring Wheeler and Joseph Dias, various hollies and the rare *Franklinia. From Massachusetts Audubon Society booklet "The Hollies of Ashumet."*

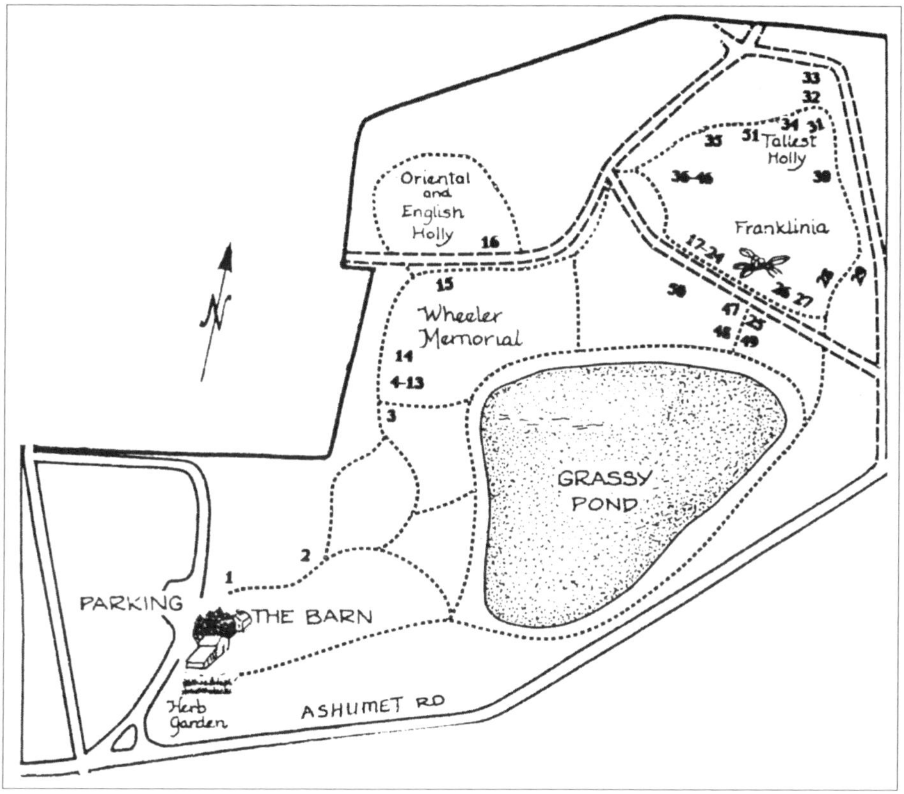

Oriental and English Holly
16
15
Wheeler Memorial
14
4-13
3
2
1
PARKING
THE BARN
Herb Garden
ASHUMET RD
GRASSY POND
33
32
34
31
51
Tallest Holly
35
Franklinia
17-24
27
47
25
48
49

was so proud, and in 1945, he proposed a Tree Planting Committee. The committee, a representative body that was charged with choosing appropriate trees for planting along roads and streets, received two successive $1,500 appropriations from town meeting.[44] During this period, Wheeler was also a member of the Town Forest Committee. He had some very specific recommendations, one of which was to get rid of wild cherry trees, which he said were weed trees, and another was to plant salt spray–resistant Japanese black pine near the sea.

By the mid-1950s, Wheeler's holly nursery was prolific. He was shipping hollies and aiding the holly business along the East Coast. Locally, besides the estate plantings he did, Wheeler had a special holly customer: Josiah K. Lilly III of Falmouth and Wellesley Hills, Massachusetts, an heir of Eli Lilly & Co. Mr. Lilly began his 2.3-acre holly farm called "Holly Acres" on the south side of Route 151 in Mashpee close to Ashumet Farm. In February 1958, Mr. Lilly paid Wheeler $2,000 for ten thousand rooted holly cuttings. Lilly's tree farm was sold in 1966.

In 1961, a delegation of the Garden Club honored Emily Wheeler at Ashumet Farm, thirty years after the first meeting of the club. She died at age ninety-seven in 1972. David Lillie felt that his aunt Emily was remarkable because she was a farmer's wife in a secluded area who was able to spin a social web that included interesting people in the arts, letters and natural science, especially the growing and use of herbs.

Wilfrid Wheeler remained on Ashumet Farm until his death at age eighty-five in 1961. Given his many years of government and civic service, you had to believe that people liked him and were comfortable around him. David Lillie said of his uncle: "He had no pretenses."

During Wilfrid Wheeler's life, the Wheeler family sold most of Ashumet Farm's original 320 acres. After Wheeler's death, his heirs sold Ashumet Farm's valued horticulture plantings of about 49 acres to local benefactor and holly enthusiast Josiah K. Lilly III, who rescued the beautiful landscape from a fate as a residential subdivision. In 1964, Mr. Lilly donated the property to the Massachusetts Audubon Society to protect the hollies and to further environmental stewardship.[45] Today, it is a Hatchville landmark known as the Ashumet Holly and Wildlife Sanctuary.

Chapter 5

ROBERT W. LEATHERBEE

BRAE BURN FARMS

Robert Leatherbee's wartime work with the War Shipping Board had removed him from the day-to-day effort to develop Coonamessett Ranch Company's model farm, which was carried out by Frederick Leatherbee and Wilfrid Wheeler. When he returned to Falmouth, Robert already had been elected president of the board of directors of the Coonamessett Ranch Company. He now began to plan his own farm, which he described in a twenty-fifth Harvard reunion publication as "experimental."

In 1921, Robert purchased land in historic Hatchville from Eugene Hatch, son of the late Silas Hatch.[46] These parcels were situated between the Coonamessett Ranch Company's farm manager's house, where the Wheelers were living, and Hatchville Road. They hugged Coonamessett Pond and extended south along Hatchville Road to encompass the old Turner homestead. This property would become the first piece of the new "Brae Burn Farms," an operation aimed at producing commercial quantities of safe milk.

To help him implement his farming plans in Hatchville, Robert Leatherbee relied on Swedish-born Hjalmar Jensen. Their relationship started around 1910, when Mr. Jensen immigrated to Chicago and found work as a machinist at the Crane Company, where Robert Leatherbee also was employed.

Leatherbee soon hired Jensen to work in the greenhouse on his gentlemen's farm, called Brae Burn, in Lake Forest, Illinois, but later, he became the family's chauffeur. While working on the farm, Jensen saved the house from burning down by extinguishing a fire in the basement. When the Leatherbees moved to Woods Hole in 1914, Jensen came too and thereafter stayed with Robert. At the Hatchville farm, Hjalmar Jensen became the trusted manager, which was especially important because Leatherbee often was absent to attend to business and family affairs in Boston.

DIVORCE AND REMARRIAGE

While Robert was planning their family's future in Falmouth, on June 14, 1921, Frances Anita deserted their marriage. Robert was granted a divorce on January 10, 1924, in a secret hearing in Waukegan, Illinois.[47]

Hoping and expecting to find stability in his life, Robert Leatherbee, forty-four, who was now leaning toward a life as a farmer, married twenty-two-year-old Helen Everhart in 1925.[48] They lived in a recently completed manor house called Lochstead, which Robert had constructed on a rise forty feet above the northwest section of Coonamessett Pond. The spacious two-story, ten-room Colonial boasted great views, a large beamed living room with two fireplaces and four bedrooms. Nearby were a garage for Robert's fast cars and a stable for horses and ponies. Robert and Helen had one child, James, born in March 1927.[49] Early in their marriage, Robert spent some time in Boston trying to get a foothold in investment banking, but eventually, he returned to Hatchville to run Brae Burn Farms.

Frances Anita's private and surprising marriage to Jan Garrigue Masaryk in December 1924 was much more high-profile news than Leatherbee's union with Helen.[50] Jan Masaryk, as a young man of twenty-one in 1907, had worked in the personnel office of the Crane Company valve factory in Bridgeport, Connecticut, managed then by Richard Crane. Masaryk was given this job to provide the young man an opportunity to experience a real work environment, compared to his usual café life in Prague. Obviously, the Cranes' connection to the Masaryks was very personal. Charles Crane was a financial and political supporter of philosophy professor Tomas Garrigue Masaryk, Jan's father, who, after World War I, was the first president of the Czech Republic. According to a wedding announcement, the couple had met in Prague in

Lochstead manor built by Robert W. Leatherbee in 1925 and later owned and renovated by Norman and Marjorie Dupee. *Courtesy Donna Jewett.*

1921 when Frances Anita Leatherbee visited her brother Richard, then the U.S. ambassador to the Czech Republic.

When Jan Masaryk represented Czechoslovakia at the court of St. James in 1925, the couple and Frances Anita's three sons moved to London, where Frances relished the social life. They spent their summer holidays in Woods Hole at White Crest. According to family members, Frances Anita tired of Jan's abusive behavior and left their marriage, which was annulled in 1931. She moved back to Woods Hole and was thenceforth known as Madame Crane. She then energetically supported her son Charles's acting troupe, the "University Players."[51] An obvious admirer described Madame Crane at this time of her life as "tall, of regal mien, very much the grand dame. Strikingly beautiful, with deep-set dark eyes, and soft, wavy graying hair, she seemed quite remote from all mundane concerns."[52] Before the end of the decade, after her father's death, Frances Anita and her brother John would be the directors of the Coonamessett Ranch Company.

BRAE BURN FARMS

According to town assessor's records, it took about a decade for Brae Burn Farms to grow from 3 cows on ten acres of pasture to 140 cows using two hundred acres. In 1922, there were more horses than cows, and Robert could ride through the Hatch homestead and watch Wilfrid Wheeler build up the model farm. A year later, Leatherbee obtained about thirty acres of land west of Wheeler's farm operation from the Coonamessett Ranch Company.[53] In 1924, there were 17 cows and one hundred fowl with hen houses, as well as a new silo, on Robert's core land on Hatchville Road. By 1925, when the Coonamessett Resort was being developed just east of his land, there were four colts, five horses, 23 cows, four calves and ninety fowl, as well as a new cowshed.

When Robert's father, Charles Leatherbee, died in 1926, Robert became the president of the Leatherbee Company lumber business in Boston and managed several trusts. These events might have given him an infusion of cash that he could use to expand his dairy farm by purchasing the nearby one-hundred-acre Tent poultry farm, giving him about four hundred acres of all types of land.[54] With the Tent property came a house, barn, pump house, henneries and woodland. Also in

Harvesting corn on a Brae Burn Farms field in 1933. *Courtesy Frederick Jensen.*

Helen Leatherbee with one of Robert's automobiles and his dog. *Courtesy Leatherbee family archives.*

James Leatherbee driving a trap at Brae Burn Farms. *Courtesy Leatherbee family archives.*

1926, Leatherbee bought the milk business of Antone Santos, called Broadmeadow Farm.

Over the next several years, Robert and Hjalmar Jensen worked to grow both the dairy and produce parts of the business. Brae Burn sold dairy products and vegetables, especially asparagus, at its stores in Falmouth and in North Falmouth. It also delivered milk, which was cooled by pond ice stored in an icehouse next to Coonamessett Pond.

Despite the hard work necessary to prosper in farming, Robert seemed fulfilled. In February 1931, when his farm was probably operating at capacity, he wrote to his wife, Helen: "Never can I adequately show or express my appreciation for what you have done for me or mean to me."[55]

FREDERICK K. LEATHERBEE: KENSINGTON FARMS

In 1929, Frederick Leatherbee returned to Hatchville from Port Chester, New York. He soon purchased about sixteen acres from Allen B. Hatch on the north edge of Deep Pond,[56] and he built a new dairy operation, Kensington Farms, on the former Randall Farm on Hatchville Road.[57] At that time, his farm and his brother's Brae Burn Farms were the major modern local milk and butterfat suppliers, as Atamannsit Farm's Guernsey stock was being sold. Since the mid-1920s, Cape Cod milk cows had been known for their high butterfat content. Barnstable County was the first in Massachusetts to have all cows tuberculin tested and the only county to have a local health official. It was thought that with these safeguards in place, Cape Cod dairies would have an advantage in selling milk products in Barnstable County despite competition from larger off-Cape milk producers.

However, by January 1934, a year after celebrating their son John's graduation from Harvard and the young man's marriage to Helen Luscombe of Woods Hole, Fred Leatherbee's Kensington Farms, Inc., filed for bankruptcy, and a month later, its assets were sold to Robert Leatherbee for $5,000. Frederick Leatherbee died at age fifty-six in September 1941 on his Vermont farm.

ROBERT W. LEATHERBEE: LIFE AND DEATH

In 1930, as president of the Cape Cod Milk Producer's Association, Robert Leatherbee had been working extremely hard to market milk and other dairy products. In his spare time, he also promoted the town of Falmouth. He rarely had time to ply the local waters on his 110-foot yacht, the *Brae Burn*, a converted World War I submarine chaser. During one summer, the yacht was a dormitory for some members of his son Charles Leatherbee's troupe of University Players. Robert was also busy as director of the Cape Cod Farm Bureau, a fundraiser for the Cape Cod Council of Boy Scouts, president of Falmouth Rotary and a member of the Falmouth Planning Board and Finance Committee.[58]

Leatherbee suffered an interruption of his farming operation in 1932, when a barn he was leasing on Coonamessett Ranch Company land burned after hay caught fire. Mrs. Alma LaBonte, a neighbor on Ranch Road, raised the alarm. Although the loss was partially covered by insurance, the fire threatened a prized bull named Horatius and a cow. Fortunately, they were rescued by John Wray, a herdsman, who suffered burns on his back and shoulders.[59] The fire burned an electric transformer, so the cows had to be milked by hand, and water pumps were stilled for about four hours.

In 1933, Robert Leatherbee's personal real estate was assessed at $95,150.00 and taxes were $2,163.38; Brae Burn Farms was valued at $19,600.00 and taxed at $441.00. At this time, he had 140 cows. For Helen E. Leatherbee, a highlight of the first months of 1934 was being elected president of the Falmouth Garden Club. The last part of the year would be a nightmare.

Before he could consolidate Kensington Farms into his dairy operation, in August 1934, Robert Leatherbee died of a heart attack at age fifty-three. Dr. Lombard C. Jones, Helen's stepfather, was present at the end of Robert's life. A timely editorial in the *Falmouth Enterprise* stated, "Robert W. Leatherbee was conspicuously a business man with a high sense of public responsibility."[60]

After petitioning the probate court for approval, Brae Burn Farms operated for a few years until H.P. Hood & Sons purchased its retail stores. In May 1936, Brae Burn Farms merged with Cape Cod Creamery.

Probate of Leatherbee's estate showed personal property of $78,438.18 and realty of $60,500.00. His shares of Coonamessett Ranch Company stock were deemed worthless. Under terms of Leatherbee's will, his wife, Helen, and the State Street Trust were named as joint executors and direct bequests and trusts were established for his sons.[61]

ROBERT LEATHERBEE'S LAND SOLD

After his death, Robert Leatherbee's wife, Helen, and the State Street Trust sold his real property, some through negotiations with separate buyers and the remainder at an auction held on August 31, 1937. Leatherbee had accumulated about four hundred acres, some still mortgaged. Amazingly, and also improbably, large pieces of this real estate are still relatively intact.

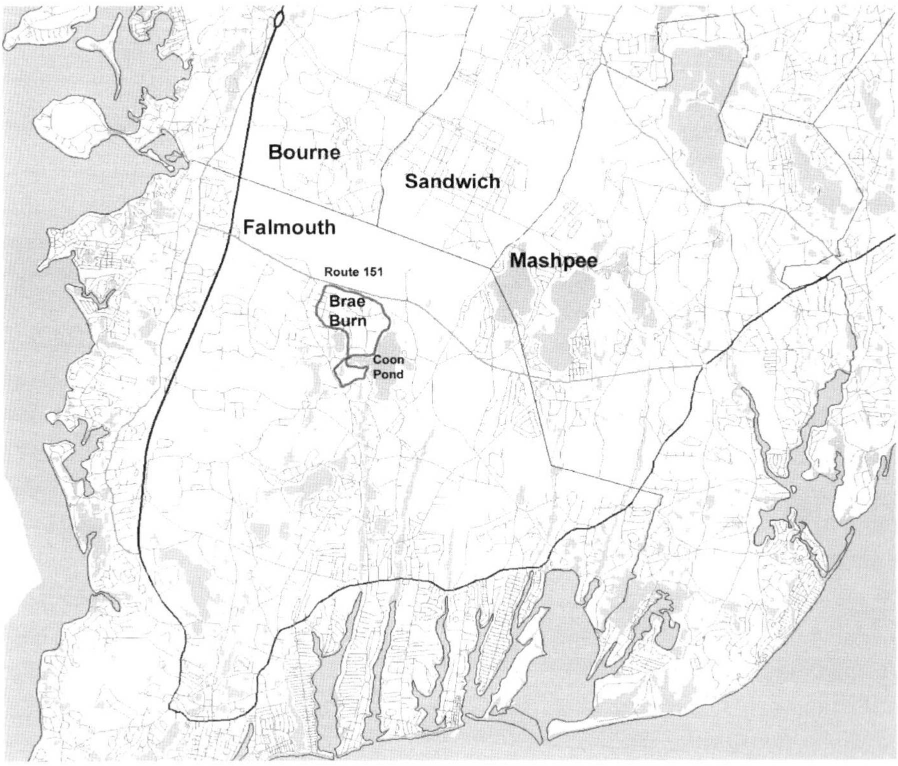

The accumulated four hundred acres of Robert Leatherbee's Brae Burn Farms. *Produced by the author from a map courtesy of Tom Stone.*

DIRECT SALES OF LEATHERBEE LAND

In addition to the sales that follow, small parcels on Deep Pond were sold directly to Ralph Hunt, Samuel Temple and Nathan Ellis.[62]

Hjalmar Jensen Buys Heart of Brae Burn Farms

The largest segment of Leatherbee's estate, parcels totaling 121 acres centered on Hatchville and Boxberry Hill Roads, was sold directly in 1936 to Hjalmar Jensen, Leatherbee's trusted manager, who had been employed by him for over twenty years.[63] Essentially, this was the same land that Leatherbee had purchased from George Tent in 1926. Most of the land was north of Boxberry Hill Road but about 20 acres were south of the road adjacent to a peat bog. Jensen later leased 100 acres west of the Coonamessett airport from the Coonamessett Ranch Company. He operated a dairy and truck farm until 1953 (see Chapter 8). Just over half of this original purchase is still intact and owned and lived on by the Jensen family. None of it is in agriculture.

Norman Dupee Buys Lochstead Property

Norman Dupee—from Waban, Massachusetts, a wool merchant of the firm of Forte, Dupee, Sawyer & Company in Boston, Massachusetts—purchased a 76.7-acre parcel, including the Lochstead manor house on the north shore of Coonamessett Pond west of the Coonamessett Inn, from the Leatherbee estate for $20,000 for a family compound (see Chapter 8).[64]

Norman Dupee Buys Former Atamannsit Farm Property

Although not part of the Leatherbee estate, the purchase of the former Atamannsit Farm is important because a good portion of this acreage has also remained largely intact to this day. In 1938, Norman E. Dupee purchased a 150-acre property along the southwest shore of Coonamessett Pond that included the Atamannsit Farm, previously owned by George St. Amant, at foreclosure from the Roxbury Institution for Savings.[65] At the time, the property included a bungalow (lodge), a hunting camp and the cow barns, which Dupee eventually demolished (see Chapter 8).

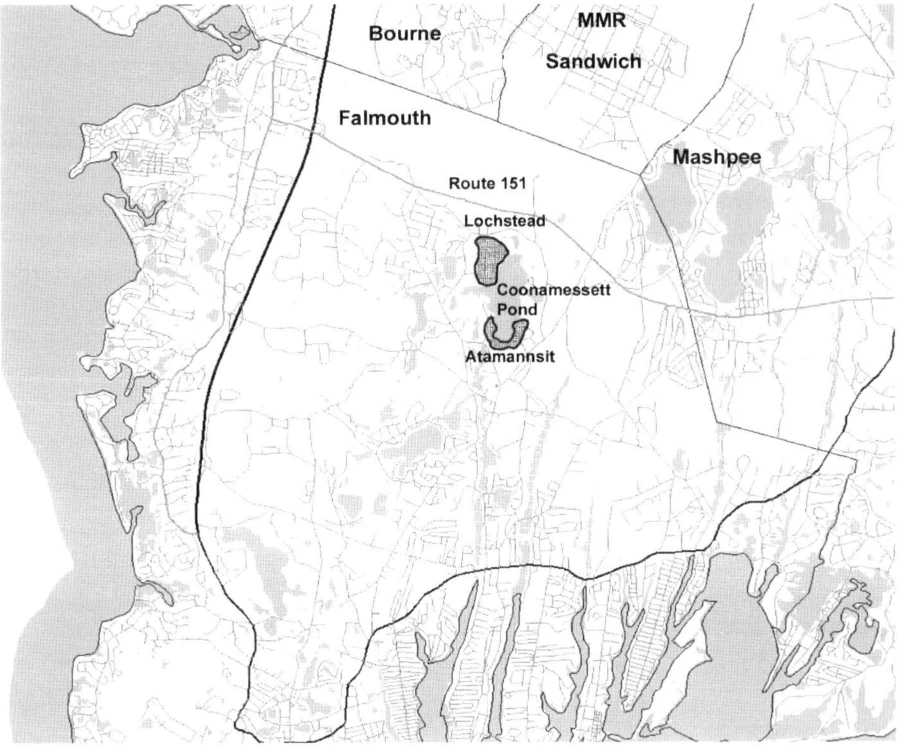

Lochstead and Atamannsit lands owned by Norman Dupee from 1937 to 1956. *Produced by the author from a map courtesy of Tom Stone.*

AUCTION SALES OF LEATHERBEE LAND, AUGUST 31, 1937

Peter Gomes, Strawberry Grower and Dealer

Peter Gomes, a Cape Verdean strawberry grower, agent and trucker from Wareham, Massachusetts, paid $550 for 102 acres west of the Jensen land extending just beyond Sam Turner Road.[66] According to Hatchville resident Fred Jensen, Gomes didn't live on the land but grew strawberries there. Twenty years later, Gomes sold his land. In the 1970s, this land was developed for residential housing.

Alma LaBonte, Poultry Farmer

Alma LaBonte, a poultry farmer who lived on Ranch Road with her husband, William, who worked for the Coonamessett Ranch Company, purchased four parcels totaling seventy-six acres and a farmhouse on Hatchville Road southeast of Crooked Pond Farm. Later, the LaBontes added a new dairy barn (from the defunct Kensington Farm) and forty acres with frontage on Coonamessett Pond.[67] To obtain this land, Alma outbid Norman Dupee. On her new farm, she likely continued to sell broilers, roasters, turkeys and ducks. In 1938, she brought in eight sheep. Her land eventually went to her daughter, Anita, and her daughter's husband, A. Robert Gunning. On some of the Gunning land, two institutions—New Alchemy Institute and Coonamessett Farm—were formed in the 1970s and 1980s, respectively (see Chapter 8).

That Brae Burn Farms and Atamannsit Farm were sold as intact parcels set the stage for a period of relative stability in land use in historic Hatchville. As yet, there was little interest in subdividing these properties. These parcels were large enough and contained sufficient residual infrastructure and soil quality for farming to continue for another twenty years on some of them nearby to other small farms and the bustling Coonamessett Resort.

PART II

AGRARIAN LIFE NEAR A RESORT ON COONAMESSETT POND

Chapter 6

NEW RESORT AND MAJOR LAND SALE (1925–1938)

The period from 1925 to 1938 saw major changes. The Roaring Twenties gave way to a deep economic recession, while in Germany, the Nazi party gradually became a dominant force that threatened all of Europe. During this time, the United States was still isolationist; nevertheless, preparations for the defense of the nation began locally with the expansion of Camp Edwards in the Massachusetts Military Reservation in the Cape Cod townships of Bourne, Falmouth and Sandwich. About two-thirds of the Coonamessett Ranch Company land would be needed to build the base to train thousands of soldiers of the Twenty-sixth Yankee Division of the National Guard. Falmouth's population grew from 3,500 in 1920 to 6,878 in 1940. Soon, an influx of military personnel and money would change the town.

In historic Hatchville, major changes occurred, too. Wilfrid Wheeler started Ashumet Farm and helped commercialize the American holly while Robert Leatherbee aggregated land and other farms to make Brae Burn a competitive dairy supplier. While these farms were flourishing, a Crane-owned resort centered on Coonamessett Pond replaced the model farm.

FROM MODERN FARM TO INLAND RESORT

When Robert Leatherbee and Frances Anita Leatherbee were divorced in 1924, Charles Crane no longer had to consider financially supporting

Robert, especially his business interests. At this time, the Coonamessett Ranch Company's model farm was not even covering its costs. Crane had to decide what to do with the usable 900 out of approximately 2,500 acres of Coonamessett Ranch Company land in Hatchville. For this decision, Crane relied on the advice of Donald Brodie, his financial secretary, and Lawrason Riggs Jr., his attorney. Subsequent actions suggest that they advised him to cease subsidizing the farm and convert the land into something potentially profitable, like a resort. Undoubtedly, more of his money would be needed for this change in land use. Crane agreed with his advisors, as Brodie replaced Wheeler as treasurer and Riggs became the Coonamessett Ranch Company's president, replacing Robert Leatherbee.

During the shift to a summer resort, the farm still supplied produce, dairy and poultry to the company stores. Eventually, the farm buildings were removed and replaced by tennis courts and various apartments. When the transition from farm to an inland resort was complete, it contributed to the 28 percent farm loss on Cape Cod between 1920 and 1930.

DEVELOPMENT OF THE COONAMESSETT RESORT

Emblematic of this change was the fate of the renovated Chadwick farmhouse that the Wheelers had occupied. In 1924, it became the "Coonamessett Tea Room" run by Perley Jordan, who had worked as a vegetable gardener for Wheeler on the model farm. The tearoom later became the core of the Coonamessett Inn, which by July 1925 had dining facilities and ten overnight rooms. The white clapboard inn, which would mushroom into a sprawling complex, was at the round end of a horseshoe-shaped driveway along which visitors could park their automobiles. Later, in May 1936, the "ranch house" addition to the Coonamessett Inn was opened. Each apartment had a private entrance, a living room with a fireplace and a twin bedroom with bath.

During this time of nationwide golf mania, it was necessary for the young Coonamessett Resort to have a challenging and yet fair public course. In 1928, along the eastern high ground on Boxberry Hill Road, a nine-hole golf course opened and was expanded to eighteen holes. The greens fee was a reasonable $1.50 when the full eighteen-hole course opened on July 1, 1930, and there was a new clubhouse, which had a lounge and showers. A new road was cut from Boxberry Hill Road to the clubhouse, which a decade later would also be the entrance to the parking lot for the Coonamessett

Shady, enticing entrance to the Coonamessett Inn. *Courtesy Daniel Small.*

Bird's-eye view of the Coonamessett Inn on the pond. The golf course is to the east (left), the beach to the south, the bog close to the west and the Lochstead garden to the far west. *Courtesy Gareth Jennings.*

Coonamessett airport, established around 1929, had grass runways and no control tower. *Courtesy Leatherbee family archives.*

Club and its successor, the Falmouth Playhouse.[68] By 1933, for $0.40, one could watch 4:00 p.m. Sunday polo matches at the only active polo grounds on Cape Cod.

To provide improved access to the inn and other buildings, Ranch Road was cut between Boxberry Hill Road and Coonamessett Road (later the Nathan Ellis Highway/State Highway 151). Opposite Ranch Road on the north side of Coonamessett Road, the Coonamessett Ranch Company developed the small Coonamessett airport. It had a grass runway, fuel depot and an office but no control tower. By this time, relatively affordable high-powered aluminum monoplanes were available for public sale. In addition, the resort hosted seaplanes on the Coonamessett Pond that could carry visitors to and from Cape Cod and offshore island destinations. The Town of Falmouth leased the airport in 1930 and 1931, and it grew steadily under the management of George M. Cluett II and Guy A. Hamm Jr.

EDNA SPRAGUE HARRIS, RESORT MANAGER

Running a successful full-service resort obviously required trustworthy and knowledgeable staff under the guidance of a smart, detail-oriented manager. The Coonamessett Ranch Company was indeed fortunate to lease

One of three cozy sitting rooms of Coonamessett Inn furnished with antiques and decorated with curios and Sandwich glass. *Courtesy Gareth Jennings.*

the operation of the resort in 1930 to Edna Sprague Harris, owner-operator of the Megansett Tea Room in North Falmouth. Edna's lease of the Coonamessett Inn was to run for at least fifteen years. Born in Watertown, Massachusetts, she came to Falmouth when her father, Chester Sprague, developed the Megansett section of North Falmouth.

An acknowledged perfectionist, Edna, as she was affectionately known, supervised more than one hundred employees, all dedicated to her and to the success of the resort. The inn's kitchen was stocked with local dairy products, chickens, turkeys and garden vegetables. She ran an antique business, which was reflected in the tasteful dinning room tableware and the inn's furnishings. Food delivery and room service were efficient and the food excellent. The resort's advertised rates were five to eight dollars per day on the American plan (room and meals) with weekly rates available.

1931: A Pivotal Year for the Cranes

While the Coonamessett Resort was being developed, external events intruded on Charles Crane's idyllic life of travel, which alternated among periods of rest at the Carlsbad Spa (*Karlovy Vary*) in Czechoslovakia and his homes in Woods Hole and Palm Springs, California. In September 1931, Richard T. Crane Jr. left "Castle Hill," a Stuart-style mansion on his 1,500-acre Ipswich, Massachusetts estate, and sailed his yacht to Woods Hole to visit his brother, Charles. The pair discussed the state of the Crane Company, which Richard had run well for seventeen years. However, Richard, who was once worth $50 million and reputed to be the second-wealthiest man in Chicago after Sears & Roebuck chairman Julius Rosenwald, was uneasy—and for good reason.

As the economic depression took hold worldwide, demand for Crane Company manufactured plumbing, valves and other metal products declined so that the income statement, which boasted a profit of $11.5 million in 1929, was showing an $8 million loss in 1930. The Cranes and other bondholders were affected by the market's downturn, but the employees were buffered somewhat by the $12 million in stock Richard had given to their pension fund and by a special fund Charles and Richard had established in 1914 to help families in need. The Crane Company always took care of its employees. Stress, however, likely took its leader. Within two months of this meeting, Richard Teller Crane Jr. died at age fifty-eight of the effects of heart disease.

Military Training Expands on Cape Cod

Coonamessett Ranch Company's unimproved land became the target for the expansion of National Guard training, which had taken place since 1908 in a forest in Sandwich. After World War I, military planners funded an expansion on Upper Cape Cod (Falmouth, Sandwich and Bourne) training facilities for the Twenty-sixth Division of the Massachusetts National Guard. This expansion was not universally embraced on Cape Cod. Nevertheless, in 1935, just after the widened Cape Cod Canal and the vehicle-carrying bridges over it opened, the Coonamessett Ranch Company sold 8,745 acres (13.5 square miles, or 62 percent of their land) of mainly woodland to the Commonwealth of Massachusetts, reputedly for $50,000. The sale included 4,307 acres in Bourne, 3,841

Camp Edwards was a National Guard training facility until leased by the federal government in 1940 to serve as a war mobilization facility for thirty thousand troops. *Public domain.*

acres in Sandwich and 597 acres in Falmouth. This land included most of what Wilfrid Wheeler had purchased between 1914 and 1917 at about $5 per acre.[69]

The Coonamessett Ranch Company bondholders at the time of the sale of this land were Charles R. Crane, Frances Anita Crane and John Oliver Crane, and the company officers were Lawrason Riggs Jr., president, and Donald M. Brodie, treasurer. Given Charles Crane's long-standing support of the Marine Biological Laboratory (MBL), it was not surprising that in 1942, Donald Brodie was elected treasurer at the MBL, replacing Riggs, who succeeded Frank R. Lillie as president of the MBL Corporation. Both men served in their new MBL posts for a decade.

Camp Edwards, which was dedicated in July 1938 before fifty thousand on-lookers, was about 15,000 acres in area within the much larger Massachusetts Military Reservation (MMR). As the federal government took over the operation, twenty-five thousand National Guard trainees populated Camp Edwards. Camp Edwards was enlarged by 6,457 acres in 1941.

Camp Edwards provided a mighty economic boost for Upper Cape Cod, especially Falmouth. First, there was the payroll for constructing the base buildings and roads, which peaked at about $1 million per week. Then, when the soldiers came to the base, they, too, spent money in the surrounding communities. Officers' families were billeted off base, mostly in rooms to let, which provided income for families.

Chapter 7
COONAMESSETT RESORT (1939–1954)

Charles R. Crane died on February 14, 1939. Cornelia Crane received many condolence notes, including one from President Franklin D. Roosevelt, which read, "I am shocked and saddened by the news of the passing of your devoted husband who was my staunch friend through many years. As citizen and man of affairs he rendered public service of great usefulness and distinction in many fields. Mrs. Roosevelt joins me in sending heartfelt sympathy to you and to all of the family."[70]

After their father's death, Frances Anita Crane, fifty-one, and John O. Crane, thirty-nine, assumed control of the Coonamessett Ranch Company and its major asset, the resort. What would they do? War was on the horizon, but the nearby military training facilities provided some cause for business optimism.

Personally, Frances Anita Crane needed a change after a horrible decade. In 1931, her marriage to Jan Masaryk dissolved and her uncle Richard died. In 1932, she had a gall bladder operation in Budapest while traveling with her father and returned to Woods Hole to recuperate. Her first husband, Robert Leatherbee, died in 1934, and her talented twenty-seven-year-old son, Charles, an actor and director, died of pneumonia a year later. Her brother Richard died in 1938. What was next for her and her other brother, John?

John Oliver Crane and his father shared many experiences and interests. They were both adventurous, entrepreneurial and enjoyed travel and living abroad. Prior to his senior year at Harvard University, John accompanied

Elderly Charles R. Crane. *Public domain.*

his father, who was finishing a stint as U.S. ambassador to China, on a trip by private rail car from China through Siberia to Europe. After he graduated, in 1922, John was invited to become private secretary to Czech president Tomas Garrigue Masaryk in Prague, a position John held for about seven years.

In October 1929, in Rome, John O. Crane married an Italian countess, Teresa Martini Marescotti, described in the *Falmouth Enterprise* as "an Italian society girl." The Marescottis and the Vatican had a special relationship, and the Crane-Marescotti wedding was held in the private chapel of Cardinal Ragonesi, a friend of the bride's parents. They also arranged for a papal audience with Pope Pius XI for John's mother, Catholic convert Cornelia Crane. John and Teresa had a daughter and later divorced.

According to Sylvia E. Crane, John's second wife, Charles Crane purchased Count Martini Marescotti's farm near Rome in 1931 for $450,000.[71] Looking back on those days, John Crane, who spoke Italian and lived in Rome as a legal resident at the time, maintained that he was able to transform the four hundred acres into a dairy farm where he "doubled or trebled production of both milk and grain."[72] Meanwhile, he honed those skills he would need to guide and expand the Coonamessett Resort.

CRANE SIBLINGS MANAGE THE RESORT

The siblings formed a partnership called the Coonamessett Associates; John was president, and Frances Anita was the treasurer. For the next eighteen years, there were two management trends: to grow the resort to make it more attractive and financially successful and to sell it.

Initially, John shuttled between Rome and Woods Hole, during which time he made some tough decisions. In 1939, contrary to recommendations from the Town of Falmouth, which had invested heavily in improving the Coonamessett airport for a municipal facility, he chose to keep the airfield private.

Transportation to Hatchville improved in June 1940 with the completion of a new three-lane road between North Falmouth and Sandwich Road, the Nathan Ellis Highway, named for the Hatchville resident and Falmouth's road superintendent.

With the earlier expansion of the Massachusetts Military Reservation, recreation for both enlisted personnel and officers became an important business opportunity for the Coonamessett Resort, just two miles from Camp

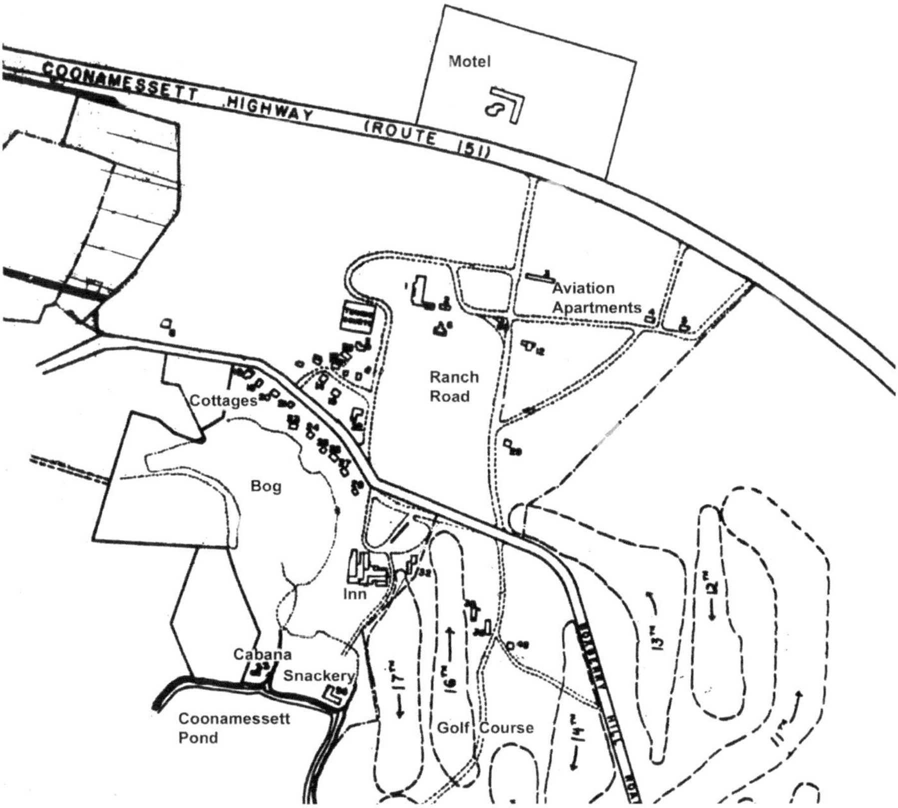

Map of the Coonamessett Resort (annotated by the author), depicting the inn, cottages, tennis courts, cabana and golf course. *Courtesy Harvey G. Clauson Jr.*

Edwards. The resort invited golfers to take advantage of one-dollar green fees during the winter. By November 1940, every man in the district was issued a draft number. Even though war was on the horizon, local tourism carried on. The Coonamessett Inn, operating year round, by then had sixty rooms with daily rates of six to ten dollars and weekly rates of forty to sixty-five dollars.

To satisfy the demand for off-base officer housing, in February 1941, the Coonamessett Associates built twenty-five prefabricated, subfloor heated cottages along both sides of Boxberry Hill Road west of the inn; ten of them looked south to a large bog and to Coonamessett Pond. These cottages were not for sale. Instead, military officers from Camp Edwards rented these colorfully painted dwellings for sixty dollars per month for a single and seventy-five dollars for a double.

Aerial view of the barn-like Coonamessett Club, built in 1941, adjacent to the golf course. *Courtesy Gareth Jennings.*

More targeted resort expansion came in the summer of 1941. The Coonamessett Club was built across from the golf clubhouse immediately north of the eleven-acre Round Pond with a stunning view of Coonamessett Pond and with road access from Boxberry Hill Road. The club, which opened on June 20, was an 80- by 120-foot weathered replica of a New England barn replete with dance floor/skating rink, bowling alley, game room, dining room and lounge. Importantly, it had a liquor license.

Club members and invited guests attended a gala opening reception. Some invitees, including the architect, contractor and three Falmouth selectmen and their wives, had dinner with Frances Anita Crane at White Crest in Woods Hole. The Cranes, of course, welcomed military brass from Camp Edwards. For example, the chief of intelligence and the commandant appeared at the club's opening night. And during that summer, John O. Crane and several military chiefs stood on the club's Sunset Terrace and viewed bomb-dropping contests over Coonamessett Pond.

Club members were expected to use the resort facilities for golf, tennis, swimming, boating, fishing and hunting, as well as dine in the solitude of rural Falmouth—sort of like a country club with a lot more activities. The club adopted a Native American logo to mirror the Algonquian *Coonamessett*, which is said to mean "place where large fish are found."

The popular tennis courts, located west of Ranch Road, were usually full with players and instructors. Club members and military officers received reduced seasonal memberships. Between Ranch Road and the tennis courts were the studio and two cottages. On the east side of Ranch Road were three cottages and the so-called aviation apartments, presumably rented to pilots who used the resort's airport.

In its first year (1941), the Coonamessett Club remained open in the fall, taking advantage of the usually dry, mild weather. It held a Camp Edwards Sports Day, a Halloween party and an Armistice Night Dance (November 11), as well as fishing and hunting parties.

Meanwhile, on November 17, 1941, just before the United States entered World War II, Cornelia Workman Smith Crane, who had suffered many illnesses during her life, died at age seventy-nine in Woods Hole.

WARTIME

Even though the country was at war, the Coonamessett Club remained open for dancing, dining and indoor sports, and in 1942, it hosted exhibitions of bowling, badminton and table and paddle tennis. The appeal was to stay at the inn and walk to the sports events.

Three of Mrs. Crane's grandsons saw action in the war. Frances Anita's son Robert served in the army air corp in England and France, was wounded in battle, returned to Falmouth and helped run the club. His brother Richard, who had been a pilot since the age of thirteen, commanded a heavy bombardment squadron at Fort Dix, New Jersey, and antisubmarine patrols at nearby Otis Field on the MMR before going abroad with the army air force. Charles Bradley, son of Josephine and Harold Bradley, served with the U.S. Army Mountain Warfare Troops.

With war raging in Europe, Frances Anita Crane could not travel to her favorite foreign haunts. Instead, she remained in Woods Hole. Knowing that the Cranes were respected in Falmouth but also that it was important to participate in town affairs, Frances became a resident voter in 1942.[73]

In February 1942, the Coonamessett Associates sought and was granted by the Falmouth Planning Board a zoning change for the Coonamessett Club from single residence to business district. The Falmouth Finance Board and Mr. Norman Dupee, an abutter, opposed this blatant example of spot, or special, zoning.[74] Today, the site where the club and its successor, the

Richard (Leatherbee) Crane demonstrating his flying prowess. *Courtesy Leatherbee family archives.*

Falmouth Playhouse, once stood is a practice green for the golf course and is zoned as business.

The Cranes had to find a way to promote recreational tourism during the war. In March 1942, they hired Arthur E. Thorsen as head of all recreational attractions. Thorsen had worked at Yosemite National Park, which was a magnet for tourists—now came a real test. Thorsen thought that people would adapt to the war measures, such as rationing of food items, rubber and fuel, and continue to vacation. He probably didn't expect the nightly blackouts in Atlantic coastal regions as the military was concerned about submarine attacks and even an invasion from the air and sea. Moreover, in the fall of 1942 on Cape Cod, most civilians were prohibited from entering Camp Edwards, some dunes and marshes in Sandwich and other coastal zones. Realistically, Thorsen recognized that he would have to make the resort attractions available to Cape residents at a reasonable price.

Thorsen began by attracting more military personnel from the MMR by adding a floorshow at the Coonamessett Club with no cover charge and no minimum purchase. Adding the floorshow may have been a response to competition from the Latin Quarter on Old Silver Beach in North Falmouth, which opened in 1940. In 1941, the Latin Quarter advertised a full-course dinner for $1.50, two floorshows with chorus girls and dancing with no cover charge.

Despite the heightened security on Cape Cod, in March 1942, the editors of the *Cape Cod Standard Times* still felt that tourism should be encouraged and that visitors should know that the Cape was not a good enemy target.

As part of the new war measures, civilian aviation up to forty miles from the coastline was banned, and many small airports were closed. In 1942, Coonamessett airport operator Guy A. Hamm Jr. was managing the Cape Aircraft Company and training twenty-four student pilots when he learned that the airport would close on August 15. He relocated to another field farther inland. Nevertheless, the Coastal Air Patrol (CAP) was given permission by the Falmouth selectmen and the town's aviation committee to use the airport for the duration of the war, after runway improvements were completed. Organized by volunteers in late 1941, the CAP was part of the army air corps.

From Woods Hole, where they spent most of the war years, John and Frances Anita directed changes at the resort. The club, which had operated continuously for nearly three years, was closed from February to May 1944 for repairs and redecoration. After the Allied invasion of Europe in June 1944, attendance at the club declined, but the inn and cottages were full. Still, John Crane expected to sell the resort by Christmas 1944, he wrote to a friend, but the resort's ownership did not change in that year.

IMMEDIATE POSTWAR CHANGES AT THE RESORT

In the spring of 1945, the Coonamessett Resort and Camp Edwards reached an agreement for the Coonamessett Club to be used for the convalescence and recreational needs of servicemen recovering at Camp Edwards Hospital Center. The military kicked off the program with a golf exhibition on June 20. On a daily basis, the soldiers would be bused to the club and served beer, soda and milk with sandwiches and cake, gratis.

In 1946, the Coonamessett Ranch Company (Coonamessett Associates) owned 2,523.2 acres in Hatchville. The company's land was valued at $53,421 and buildings at $213,325, which included the inn and golf course. Its tax bill was $8,109.

For the next two years, the resort's leases changed hands annually. In 1946, it was leased to Robert B. Kretshmar, who introduced the "New Coonamessett Club," where dinner and dancing could be enjoyed for a minimum of $1.50 on weekdays and $2.00 on Friday, Saturday and holidays. The Coonamessett Club in 1947 opened under new management as "Les Crane's the Coonamessett," featuring "Caper Time" from 3:00 p.m. to 6:00 p.m. daily, at which time cocktails and hors d'oeuvres sold for $0.40—truly a very happy hour. Lester T. Crane (no relation to the owners), a Falmouth Main Street nightclub operator, proposed to increase inn and club business by catering to large groups of up to four hundred by attracting companies holding golf outings for management employees.

COONAMESSETT AIRPORT

After the war, the small airport became a political football. The airport was idle after the CAP left in 1944. In 1946, it was leased for a year to Robert L. Kincart, to serve the needs of the Coonamessett Club, which meant restricted flying. However, much work would be required to obtain Civil Aeronautics Board approval for the field's operation. Despite the lease, Kincart was denied a permit to operate the field by the Falmouth Board of Appeals for three reasons: because of the proximity to Otis field on the MMR operated by the U.S. Navy; because the airport was in a single-residence district, not a business district; and because the public would not be substantially served.[75]

Needing an active municipal airport not encumbered by restrictions imposed by the navy based at the MMR, the Town of Falmouth cancelled its lease arrangement with the Coonamessett Ranch Company and, in July 1946, moved to the new airport in Teaticket, a few miles to the east. The Teaticket airport advertised charter flights, flight instruction, airplane rentals and repairs.

Within a year, the Coonamessett airport had new managers. George Pinto of Hatchville and Robert W. Draper of Menauhant vowed to make the airport the best on Cape Cod. They began with a flight training school in August 1947. At the time, the field was equipped for night use, and planes could take off and land from any compass direction. In August 1948, members of National Guard and officers' reserve corps units received flight training there. Sadly, a Halloween air show followed by a dinner dance at the Coonamessett Inn was marred by the death of a parachutist.[76] Within a decade, Pinto would begin a struggle to retain his lease on the airport.

INTENT TO SELL RESORT

A real estate advertisement from the *Boston Herald* of February 9, 1948, which did not identify the seller, shows the extent of the Coonamessett Resort. The Ranch Company's attorney, however, confirmed that the Cranes were ready to sell the resort for a price:

> *Cape Cod's Most Famous Resort Colony*
> *Now offered for sale to individual or syndicate. Property includes: An Inn with 18 guest cottages, club house with night club and room for summer theatre, private airport and administration building, nearly a mile of lakefront, two sand beaches, 18 hole golf course, tennis courts, riding stable. Represents investment of three-quarters of a million dollars. SUBSTANTIAL CASH REQUIRED FOR PURCHASE.*

COONAMESSETT CLUB BECOMES FALMOUTH PLAYHOUSE

With no buyer for the resort in 1948, John and Frances Anita Crane continued to expand, adapt and seek outside investors. In the fall of 1948, negotiations with producer/director Richard Aldrich led to the conversion of the Coonamessett Club into an air-conditioned, six-hundred-seat summer theater—the Falmouth Playhouse—at an estimated cost of $40,000. For the comfort and pleasure of the patrons, there was a glassed-in supper club, a soft drink bar, candy counter and a women's dress shop. In the theater, comfortable maroon plush seats had good lines of sight to the stage, and the lighting and production equipment were first rate. Sprinklers connected to a fire alarm system and to fire stations that were not exactly nearby.[77]

Aldrich, who also leased the Cape Playhouse nearby in Dennis, agreed to a nine-week schedule featuring Hollywood stars performing in recent successful Broadway plays. Bookings were shared between the two playhouses. With actors such as Henry Fonda, Helen Hayes and Claudette Colbert living and working at the resort, there was definitely plenty to talk about in Falmouth. Actors could be seen in the resort's bar and restaurants, glimpsed around the Star Cottage overlooking Coonamessett Pond and spied golfing.

Quickly, the actors hit the boards. The season opened on time with Tallulah Bankhead in Noel Coward's comedy *Private Lives*. Initially reluctant to perform in a summer theater, Ms. Bankhead changed her tune when it became apparent that she could "knock off as much as $5,000 a week from June to September in roles I could have played in my sleep."[78] In fact, the first week grossed $13,400, a new record for a summer theater. Between rehearsals, the actors and staff could use the resort's facilities and dine in a separate dining room at the inn.

Aldrich was the producer through the 1952 season at the same time as he was developing musical productions at the Cape Cod Music Circus (Melody Tent) in Hyannis, Massachusetts. After the death of his wife, Gertrude Lawrence, in 1952, Aldrich sold his interest in the Melody Tent and pursued a career in diplomacy. For the 1953 season at the playhouse, the very able Sara "Sidney" Gordon produced the shows, following the same star formula.

In the fall of 1949, Frances Anita Crane sold White Crest, her home on Buzzards Bay, to Samuel Anderson Peck of Westbury, New York, for $60,000. His wife, Barbara Peck, was the daughter of Gertrude Vanderbilt and the very wealthy Harry Payne Whitney.[79]

SKYLARK MOTEL ON ROUTE 151

The last expansion of the resort by the Cranes was the 1952 construction of a motel opposite Ranch Road on the north side of Route 151. Robert Crane (who, like his mother, reverted to the Crane surname from Leatherbee) supervised its design and construction. He named the motel "Skylark" after a play of the same name first staged at the Cape Playhouse in Dennis, Massachusetts, in 1939, starring Gertrude Lawrence. A windstorm on January 30, 1953, collapsed a section of the nearly completed motel, making Robert Crane cut short a vacation in the West Indies to return to oversee the reconstruction. The motel still opened that summer. Serving the airport and

Skylark Motel located on the north side of Route 151. Built by Coonamessett Associates in 1953, it soon became the Treadway Lodge before it was sold to Harvey G. Clauson Sr. *Courtesy Gareth Jennings.*

the resort, the motel advertised as the "Skylark Air Motel" with amenities such as a swimming pool, roof top solarium, snackery and club lounge with a pianist.

EDNA S. HARRIS, UNCERTAINTY AND DEPARTURE

For fifteen years, Edna Harris had run the Coonamessett Inn profitably and without any apparent glitches. She had a lease that would expire in 1946, but with uncertainty about a change in resort ownership, in 1945 Edna opened a new restaurant, Coonamessett Corner in North Falmouth, where she employed some high school students from Hatchville farm families.[80] Her lease to run the inn was renewed.

Edna had a loyal following and, thus, supposedly had a good bargaining position with the Cranes. Every year, she served many of the same families and guests. This situation continued until 1953, when she was denied the opportunity to renew her lease to run the resort. Instead, the Treadway

From 1930 to 1953, Edna S. Harris leased the Coonamessett Inn and related facilities from the Coonamessett Ranch Company. She opened the Coonamessett Inn in Falmouth in 1953. *Courtesy of the Falmouth Historical Society, Falmouth, Massachusetts.*

Hotel Company replaced her. Edna later recalled that it was "quite a jolt. All the years I'd loved the place and held it close—then suddenly without warning I was empty handed."[81]

Edna Harris, then age sixty-four, didn't retire. She was a widow with four children and needed and wanted to work. Without much delay, she took the Coonamessett Inn name, its liquor license and some staff and furniture and secured funding from some wealthy friends. Within five weeks, the new Coonamessett Inn opened at the northeast corner of Jones Road and Gifford Streets in Falmouth, where it is today.

TREADWAY HOTEL COMPANY MANAGEMENT

The Treadway Hotel Company assumed the lease for the operation of the Coonamessett Resort in 1953, and thus the Coonamessett Inn became the Treadway Inn and the Skylark Motel became the Treadway Lodge. J. Frank Birdsall Jr., the innkeeper, leveraged the adjacent recreation facilities so that the Treadway Inn soon became a popular business conference center. During the summers and the early, warm fall days, guests bathed at the

Above: Enjoying the sandy beach and cabana at the north end of Coonamessett Pond. *Author's collection.*

Left: John Oliver Crane and his second wife, Sylvia Engle, on the terrace of the Falmouth Playhouse. *Enterprise file photo*.

north end of Coonamessett Pond, where along the sandy beach there was a cabana and the "snackery" with painted pink elephants above the bar. During the Treadway period, the Coonamessett Music Society, which had been organized by John, Frances Anita and Sylvia Crane, brought classical music to the playhouse in the cooler months.

VIOLENT DEATH OF FRANCES ANITA CRANE

The Coonamessett Associates' plans were brought to an abrupt halt at approximately half past one in the morning of August 10, 1954. Frances Anita Crane, sixty-six, and a friend and sometime nurse, Juliet Wells, eighty-one, returning to Woods Hole from opening night at the Falmouth Playhouse, were killed when their automobile collided head on with a car driven by an airman stationed at Otis Air Force Base on the MMR. The accident occurred in the rain on Sandwich Road, north of Brick Kiln Road. Tallulah Bankhead, who had starred in *Dear Charles* the previous evening, had joined Crane and Wells at the Treadway Inn, where they listened to the pianist.[82] Madame Crane, who had only resumed driving three weeks earlier because of a painful knee, was at the wheel of a friend's car.[83]

In the presence of his attorney, Robert J. Riley, twenty-two, of Cambridge, Massachusetts, and a member of the 564th Air Police at Otis Air Force Base told Falmouth police that he was in the right-hand lane (going north on rain-slick Sandwich Road) when he saw a "blaze of light" from approaching head lamps. He tried to avoid the collision. He was charged and found guilty of driving to endanger, but he appealed the decision and the $100 fine.[84]

Soon after Frances Anita's death, the Coonamessett Associates was dissolved. The Boston real estate and fiduciary management firm of Minot, DeBlois and Maddison took control of the resort's leases. After Frances's death, neither her son Robert, who quit as assistant treasurer, nor her brother John had much interest in continuing to own the resort. It was time to sell and move on.[85]

Chapter 8
LIFE IN RURAL HATCHVILLE
1940 AND BEYOND

After the three large dairy farms—Brae Burn, Kensington and Atamannsit—closed, historic Hatchville changed in one respect. By 1934, gentlemen's farms were a thing of the past. Even though the land on which these three dairy farms had sat was sold, it wasn't until decades later that residential housing appeared. Norman Dupee's Atamannsit and Lochstead properties on Coonamessett Pond were summer recreation sites, whereas some of the buyers of Robert Leatherbee's other land also farmed. As a result, historic Hatchville clung to its essential agrarian roots and rural character until the mid-1950s. Hatchville was bucking the trend that showed only 8 percent of Cape Cod land was in farming.

Farmers were able to stay in business during and after World War II even though food and material rationing and shortages were the norm because they were provided with gasoline and could buy farm machinery. The Hatchville farmers sold poultry, produce and dairy to Falmouth residents and visitors via retail stores, farm stands and even door-to-door sales. Obviously, the farmers grew and raised most of their own food, which was also shared with their neighbors. Such mutual aid, typical of farming communities, contributed to a unity that defined historic Hatchville throughout this period. Even though these farmers did some business with the Coonamessett Resort and welcomed the income, most of them were not dependent on it for their livelihoods. Out of these very local human networks emerged the character of present-day Hatchville. As Carol Razinha Baker recently said, "It's the people that make Hatchville so special."[86] Who were those people who carried on the agrarian traditions in their neighborhood?

TENT'S POULTRY FARM AND HOPKINS FAMILY

In 1916, George Edward Tent, a businessman from Brookline, purchased from Wallace Hatch a large farm of about one hundred acres, to raise poultry. By 1925, Tent's poultry farm had two horses and five hundred fowl, a house, a barn, a pump house, two henneries, twelve acres of tillage and one hundred acres of woodland.

A year later, Mr. Tent bought thirty-five acres of woodlot on Deep Pond from Silas Hatch. During the next several years, Tent cleared the land and erected several dwellings, including a garage and a large house with a well-tended flower garden. In 1927, Tent had a house, garage and twenty acres of land on the south end of Deep Pond.

George Tent went to live with his in-laws, Walter and Esther Hopkins, also on Deep Pond, off Hatchville Road. Between the two houses was a factory that made metal polishing cloths until it had to close when World War II interrupted its source of the raw materials. Walter Hopkins had a roofing company on Depot Avenue in Falmouth known as the "Hopkins Company," which served Falmouth Public Schools for roofing, waterproofing and painting.

George Tent's granddaughter, Margaret (Peg) Hopkins, the daughter of Elizabeth Tent and Richard Henry Hopkins, was born in Boston, grew up in their house on Deep Pond and attended public school and the Knox School in New York. She played with the Williams children at Crooked Pond Farm just down the road from her home and recalled that Edith Williams would make delicious ice cream. She also played with the Jensen children at their farm, which used to be her grandfather George's.

As a teenager, Peg had a horse, and she and Jane Landers would ride together. Peg recalled, "We would ride our horses to the blacksmith shop in Falmouth, spend the whole day, take them down to the ocean [and] then ride them back home along Sandwich Road."[87] Peg also rode her horse all through Camp Edwards.

Peg and her father, Richard, raced their Beetle Cat named *Susfu* (military slang), a twelve-foot, gaff-rigged wooden catboat modeled on the larger catboats used for fishing around Cape Cod. They were early members of

George Tent bought the Wallace Hatch homestead of about one hundred acres in 1916 to raise poultry. He sold the entire property to Robert Leatherbee in 1926. *Courtesy Margaret Hopkins Freeman.*

the Falmouth Yacht Club, which was commissioned in 1949. Peg was the first junior commodore at the Yacht Club, and she designed the club flag that is still in use today.

Richard Hopkins was a military man. He joined the National Guard 211th Regiment in 1939. During World War II, Hopkins was awarded a Bronze Star for operations in Belgium. After the war, he rose in the National Guard, retiring in 1959 as a brigadier general.

By Thanksgiving, Peg recalled, they usually had significant snowfall, and Deep Pond would freeze solid by December so that they could ice skate on

Frank Rego (standing at far right) transported Hatchville students to their new school beginning in 1922. He retired in 1966. *Courtesy Edward and Barbara Towers.*

In 1948, Peg Hopkins, who wanted to be an airline stewardess but was too tall, was chosen "Miss Coonamessett" at the Coonamessett Flying Club's harvest dance. *Courtesy Margaret Hopkins Freeman.*

it. Her father would build a bonfire so that the children could stay warm and skate all day.

When all one-room schoolhouses in Falmouth were shut down, as mentioned in Chapter 1, the pupils from Hatchville were bused to Falmouth village or to East Falmouth Elementary schools. After 1922, their school bus driver was Frank Rego, who drove a school bus for forty-four years. Conveniently, Frank did double duty as a truant officer for the school system. A master of the road, he drove his Teaticket Scout Troop to the 1939 World's Fair in New York City.

During Peg's teen years, the Coonamessett Flying Club sponsored the first of what became an annual air frolic and harvest dance at the Coonamessett Club (1948). Peg Hopkins was chosen "Miss Coonamessett." She received a watch that didn't work.

Peg accompanied her grandmother to the Coonamessett Inn to visit with Edna Harris. Much later, Peg worked at the Coonamessett Inn as a dining room hostess and was quite upset when Edna Harris left the Inn and moved to Jones Road in 1953. When Mr. Clauson bought the resort in 1957 (see Chapter 9), Peg continued her hostess job. She recalled, "We all pitched in and ran the inn. It did quite well and a lot of local people were employed there."[88] Officers from Otis Air National Guard Base on the MMR would eat lunch at the inn, and that is where she met Michael Freeman, her future husband.

Peg learned to pilot small planes at the Coonamessett airport, even though she was too young to get a pilot's license. She always wanted to be a flight attendant, but she was just too tall. Her second career choice was as a professional model, which she pursued in New York City. After Peg and Michael's children graduated from high school, Peg became a clothing designer and had her own shop for about a decade. In 1993, the Freemans moved back to Hatchville Road. She was often a patient advocate at Falmouth Hospital. Peg died in 2011.

C. MORRIS WILLIAMS: CROOKED POND FARM

C. Morris Williams, from Newton Highlands, Massachusetts, was another city dweller who wanted to raise poultry, which at this time was considered a way to make a good living. Like George Tent, he, too, chose Hatchville. Williams started Crooked Pond Farm on Sam Turner's homestead in 1924,

County extension agent Bertram Tomlinson (right) with poultry farmer C. Morris Williams at Crooked Pond Farm. *Courtesy Nancy Williams Chute.*

Nancy (left) and Mary Williams care for a heifer as their 4-H project on Crooked Pond Farm. *Courtesy Nancy Williams Chute.*

which he purchased from Reverend George Lawrence Parker. Crooked Pond Farm, the oldest farm in the village, is located on the west side of Hatchville Road before Sam Turner Road slices northwest.

Williams's farm started slowly, mainly supplying eggs and chicks but also vegetables. By 1934, Morris Williams had eight hundred fowl, several brooder houses, a hennery, a henhouse, a farmhouse, a barn, four acres of meadow and twenty-one acres of woodland on picturesque Crooked Pond.

Williams raised Rhode Island Reds, Barred Crossbreed and Plymouth Rock chicks and sold them for thirteen dollars per hundred. He had egg incubators in the basement, which ran on electricity supplied by a windmill and stored in batteries. For water, he dug a trench from Crooked Pond and, using a gasoline-powered pump, moved the water to the chicken enclosures.[89] In 1937, he also offered fresh-killed poultry, fresh eggs and vegetables.

In 1926, Williams married Edith E. Foster from Cambridge, Massachusetts, who became an active member of the Falmouth Garden Club. They had three children: Arthur S., Nancy H. and Mary. The family lived in a three-quarter Cape house. Arthur and his close friend James Leatherbee (Robert and Helen's son) were the same age. Jimmy, as he was called, would ride his pony across Hatchville Road to Crooked Pond Farm and all through Hatchville; presumably he let his friends ride too. The Williams children all had farm chores and 4-H projects, like caring for a heifer. On Sundays, the Williams and the Jensen families would alternate driving the children to Sunday school at St. Barnabas Episcopal Church on Main Street in Falmouth village.

After attending Falmouth schools, Nancy graduated from Mary C. Wheeler School in 1947 and then graduated from Colby College in 1951. She married Edward Hinckley Chute in 1953. Their reception was at the Coonamessett Inn.

Arthur attended Loomis Prep School and served as a U.S. Navy fireman in the Pacific for two years, after which he returned in 1946 to receive his Falmouth High School diploma.

The chick and egg business was booming in 1940. Baby chick sales increased twofold because the sex of the chick could be determined by mating Barred Rock hens with Rhode Island Red roosters, resulting in males with a white spot on the top of their heads. Fertile eggs were in great demand too. In March 1940, Morris Williams sold ten thousand eggs to poultry men.

During World War II, many people raised broilers for the pot, so they needed chicks, which Crooked Pond Farm supplied. The farm products at this time also included Edith's chicken potpies, vegetables, corn and strawberries. Morris also sold eggs to Betty Jensen, who ran a bed-and-breakfast in her farmhouse on the corner of Hatchville and Boxberry Hill Roads. After the war, Williams rented cottages he built along the shore of Crooked Pond and hosted workers from the Falmouth Playhouse in a farmhouse guest room.

These Hatchville farming families supported each other. For example, in May 1950, Edith Williams hosted thirty-three guests for a supper in support of the Cancer Fund. For this event, she was helped by most of the local wives. And Morris Williams, Hjalmar Jensen and David Dimmock often attended meetings of the Farm Bureau together.

After the war, farmers were faced with a dilemma. In order to compete with imported foods in the local grocery stores, they had to choose between growing their business or selling some of their land. Morris Williams expanded by developing homes in nearby Shady Lane off Hatchville Road, which became available after a foreclosure of that property. He also began to sell trees. After the hurricane of 1944, he chose Japanese black pine and had a dig-your-own operation. He later added other evergreens for Christmas tree sales. Williams was good friends with Wilfrid Wheeler of Ashumet Farm, who advised him in his tree business.

Today, to protect rare plants, there is a town-initiated conservation restriction on twenty-five acres of Crooked Pond shore and several wetlands. Nancy Williams Chute and her husband, Edward Chute, who moved back to the tree farm after C. Morris Williams died in 1986, have a conservation restriction on part of the farmland.

HJALMAR JENSEN'S DAIRY FARM

Hjalmar Jensen was an immigrant from Sweden who came to America to make a life for himself—not that his life in Sweden would have been bad, but perhaps less promising. Surprisingly, even though his family operated the largest dairy farm in Hatchville between 1937 and 1953—right at the intersection of Boxberry Hill and Hatchville Roads—he has been invisible to local historians. Maybe it was because he didn't advertise. It's amazing,

Jensen family, circa 1936 at Ranch Road cottage. *From left*: Greta, Fred, Hjalmar holding infant Henry and Betty. *Courtesy Frederick Jensen.*

though, that he has been overlooked since his story is inextricably linked to Brae Burn Farms and its well-known owner Robert Leatherbee.

When the Leatherbees moved to Woods Hole from Chicago, Hjalmar Jensen came with them and eventually became the Brae Burn Farms manager (see Chapter 5). Elizabeth (Betty) Pauline Johnson moved from a situation at the resort community of Beverly Farms, Massachusetts, and became a maid for the Leatherbees, where she met Hjalmar. Married in 1930, they honeymooned in Atlantic City, returning to a cottage on Ranch Road.

Betty and Hjalmar had three children. The eldest was Greta, who probably was named after Greta Garbo, whom Betty met as a teenager at a Stockholm theater restaurant where they both worked. Frederick was the second child, followed by Henrick (known as Henry), who was born in 1935.

When Robert W. Leatherbee died, Jensen bought parcels totaling 121 acres from his estate. Mr. Jensen soon leased 100 acres west of the Coonamessett airport from the Coonamessett Ranch Company where he grew wheat, vetch, hay, corn, oats, rye and millet.

Jensen's farm ultimately had thirty-six cows—a mix of Guernseys, Holsteins and Jerseys—which won prizes for milk excellence and butterfat production. The cows, some of which were obtained from the Brae Burn Farms herd, were pastured on land to the west and north of the farmhouse. Initially, all were milked daily by hand until the farm purchased milking machines, but when electricity was lost, especially during the winter, hand milking returned.

Mr. Jensen at first sold his milk wholesale to the Hood plant in North Falmouth, but after the plant stopped purchasing milk, he sold his milk to the L.S. White Dairy on Shore Street in Falmouth near Beach Road on the former Beebe Farm. Laurence S. White produced pasteurized milk from his own herd and purchased milk from other suppliers. White sold his dairy business in 1948 and subdivided the land to build summer cottages. Afterward, Jensen sold his milk to Mystic Lake Dairy in Marstons Mills, Massachusetts, which was owned by another Swede, Hilding S. Hord. Mystic Lake Dairy, which began operations in 1922, delivered daily to Falmouth, a round trip of about thirty-five miles.

The Jensen farm also had a dozen laying hens and a couple pigs that were killed at two years of age. The farm's strawberries were shipped to Boston. To preserve some farm produce and dairy, the Jensens rented space in the Falmouth Frozen Food Locker plant on Main Street.

Jensen's farm did well during the war years. Hjalmar had a steady helper who boarded with the family, and in the summer, when more help was required, his neighbor Peter Gomes brought in a crew of African Americans from Wareham, Massachusetts, who Fred Jensen called just "the greatest people." Prisoners of war from Camp Edwards were permitted under the Geneva Convention to work in agriculture. Some of them helped at the Jensen Farm and elsewhere, especially harvesting the valuable strawberry crop in East Falmouth.[90]

Hjalmar Jensen celebrated his sixtieth birthday in 1945 when his son Fred was thirteen years old. Fred Jensen recalled working almost every day after school, milking and feeding the cows, and every weekend doing whatever chores needed attention. In contrast, the summer work was very demanding. Typically, father and son would rise at 5:00 a.m. to care for the cows, clean the barn, plow, plant, cultivate, cut and harvest. Years later, Fred, who loved the farm work, explained the primary reason for his contributing to the farm chores: "There is a lot of work to be done. You have to work to make money on a farm. If you pay someone to work, there goes the money. Father used me in place of hired help."[91]

In May, the Jensens planted potatoes and two kinds of corn: sweet corn for human consumption and a higher-yielding, less sweet variety for the cows.

Harvested in August and September, feed corn was chopped and stored in the silo for use during the winter. The sweet corn was eaten at home and sold to stores in town, at a roadside stand and door to door for fifteen cents a dozen.

Betty Jensen operated a successful bed-and-breakfast in their farmhouse. Three guest bedrooms were filled all summer and most weekends in the fall and spring. In the summer, waitresses and waiters from the Coonamessett Inn rented rooms, as did many regular guests. The weekend rate was about twenty-five dollars, which included a bedroom, a shared bath and breakfast. As Fred Jensen recalled, "Mom served a fabulous breakfast at fifty cents to anyone, and she got lots takers from the staff at the Coonamessett Inn."[92] Another group that stayed at the Jensens' were cadets learning to fly at the Coonamessett airport, prior to its closing in August 1942.

Racial segregation during this period was not foreign to Falmouth. These Swedes, however, were race blind. Hjalmar Jensen always had several black friends and neighbors, including Peter Gomes and a man called "Strawberry Joe Mendes," who lived on their land. According to Fred Jensen, Hjalmar gave Joe a cottage and land to grow strawberries and melons. Jensen supplied all the plants and materials to Joe and then harvested the strawberries for the Boston market. And during the war, when African American noncommissioned officers had a hard time finding lodging in Falmouth, Betty Jensen rented to an African American sergeant and his wife.

In January 1949, the Jensens renovated a wing of the farmhouse to host a Latvian family, the Leipens, who were displaced by the war in Europe. They communicated with the Jensens through another displaced person who worked at the Dimmock's Hunky Dory Farm. Mr. Leipens, the father of the family, was familiar with farm machinery and helped Mr. Jensen.

Greta Jensen hardly ever worked in the barn or fields as she and her mother had many inside tasks. To save money for college, Greta worked as a waitress at Edna Harris's Coonamessett Corners restaurant in North Falmouth. By the time Greta graduated from Radcliffe College in 1953, she had married Fletcher P. Tomic, a Harvard graduate. Greta died in 2013.

Fred Jensen graduated from Lawrence High in 1950. His farm chores made it difficult for him to participate in after-school activities, but he did find time to play baseball (left field) and sing in the glee club. He, too, worked at Coonamessett Corners, as a cook's helper, and delivered milk for Hunky Dory Farm. Fred attended Worcester Junior College and the University of Massachusetts before becoming a navigator in the U.S. Air Force in 1953, a career he followed for several decades.

Greta Jensen (center) and her two brothers with Richard Dimmock of Hunky Dory Farm (far right) in 1944. *Courtesy Frederick Jensen.*

Henry Jensen's classmate, good friend and neighbor Jimmy Razinha was a member of the high school Travel Club along with Henry. Jimmy and his family worked and lived on the Dupee property across Hatchville Road from the Jensens. During his time at Lawrence High School, from which he graduated in 1953, Henry worked at the Coonamessett Corners as a bus boy and also helped with the cows at Hunky Dory Farm. During summers, while a student at the University of Massachusetts, he, too, drove a Hunky Dory milk truck.

The year that Henry graduated high school and Fred left for the air force, Hjalmar Jensen had his first stroke. With no family to run the farm, in February 1953, it was leased to the Skulsky brothers of Plymouth, who also bought Jensen's cows. Hjalmar Jensen retired, but the hay fields were still tended and the hay sold with the help of his neighbors. Mr. Jensen, then sixty-seven, still had a substantial vegetable garden and raised chickens for home use.

Henry graduated from the University of Massachusetts in 1957, after which he got a job as a statistical clerk with the U.S. Marine Fisheries in Woods Hole, which enabled him to be near his father and mother. Hjalmar

Hunky Dory Farm dairy delivery van, which was driven over the years by Fred and Henry Jensen and some Falmouth teachers. *Courtesy Ann Adams.*

Elizabeth "Betty" Jensen and Hjalmar Jensen, Easter 1956. *Courtesy Greta Jensen Tomic.*

Jensen died in 1963. Later, Betty sold several parcels of land, including the farm buildings and their house at the intersection of Hatchville and Boxberry Hill Roads. She then moved to Falmouth and finally lived at the Megansett Nursing Home in North Falmouth, where she died in her ninety-seventh year. About seventy acres of pastureland and fallow fields along Boxberry Hill Road were left for her three children to build their homes. None of it is in agriculture.

DAVID DIMMOCK'S HUNKY DORY FARM

A new farmer, David Dimmock; his wife; and son, Richard, came to Hatchville in 1939 to start Hunky Dory Farm, where they raised chickens and kept dairy cows. This Hatchville Road property, which had been originally owned by the Robinsons, is adjacent to the East End Meetinghouse cemetery. When Clinton Baldic's farm was foreclosed by the federal land bank, Dimmock acquired the land.

As a young man, Mr. Dimmock was a railroad laborer in Pocasset and had worked with his father and grandfather in the coal and wood business. In a short span during the Depression, David Dimmock's wife died, and he lost his job. Father and son then moved in with David's parents. David later managed the Washburn Grain Company store in Falmouth.

Despite the diminishing success of small dairy farms and the attendant financial risks involved, Hunky Dory Farm pushed on and sold retail milk, eggs, cream and other farm products, which were delivered within Falmouth. Dimmock hired neighbor John Araujo to work on the farm.

For David Dimmock, farming was a perpetual challenge. In 1944–45, there was a bad drought, so he sold everything but ten cows and took a job as a math teacher. Later, he built up the herd and returned to dairy farming on twenty-nine acres, even selling raw milk. He sold milk to Lawrence High School and hired other teachers to work for him. At one time, Hunky Dory Farm had five thousand laying hens and two poultry houses, but even that number was too few for a good business. Dimmock tried keeping bees but was shut down by the bee inspector. On another occasion, the health inspector told Dimmock that he couldn't spread manure because the neighbors on Sandwich Road complained about the smell. Dimmock, according to Henry Jensen, was a stubborn Yankee who really tried to be self-sufficient. Nevertheless, in April 1950 the Dimmocks employed a Polish couple, Mr.

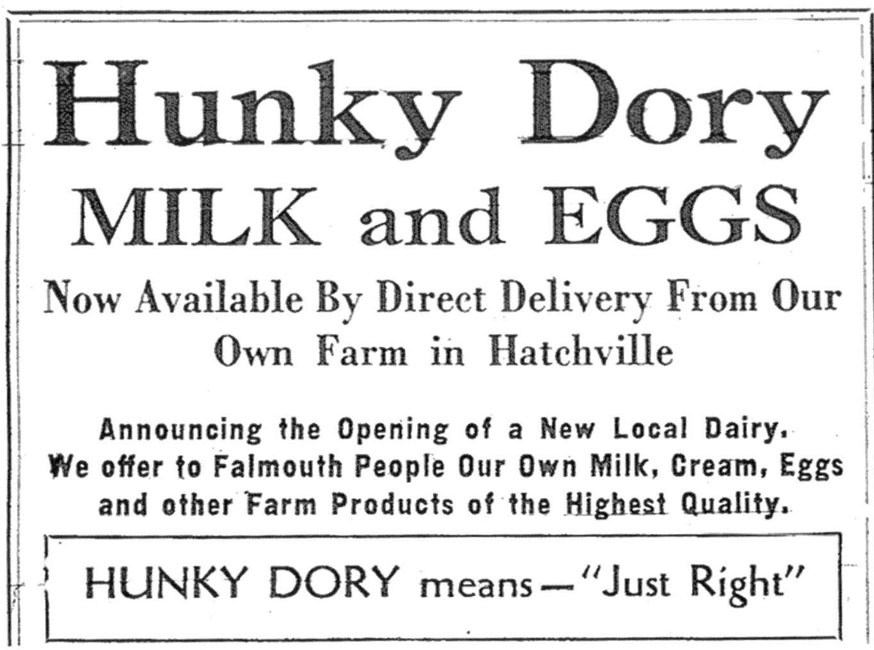

David Dimmock's 1942 announcement of a new delivery service of local dairy products and eggs. *From* Falmouth Enterprise.

and Mrs. Michael Halizcenka. David Dimmock's experience demonstrated that farmers in the 1940s and 1950s needed deep pockets in bad times and had to grow bigger to compete.

Richard Dimmock remembered Bertram Tomlinson, the Barnstable County extension agent since 1924, coming to the farm and holding a cardboard box with white powder to treat flies and fleas. Most of the ingredient was DDT. Arsenic and DDT, spread from a hopper, was the way they treated potato bugs.[93]

In 1947, David and his wife, Dorothy Dimmock, and neighbors Roy and Ethel Allen were town meeting members. Dorothy Dimmock was a member of the nominating committee the Falmouth Garden Club.

Richard Dimmock graduated from Lawrence High School in 1946. His classmate was neighbor Eugene H. Allen. Dimmock later graduated from Guilford College in North Carolina. He retired from the Woods Hole Oceanographic Institution (WHOI) after forty-three years, the last as head of marine operations. But as a high school student, he told the Lawrence High

School yearbook that his "fate was dairy farming, and that he was happiest milking cows." According to Henry Jensen, David Dimmock stopped farming about 1957. David Dimmock died in 1981 at the age of eighty.

ROY AND ETHEL ALLEN

During the Depression years, Roy Allen started a battery refurbishing business in Falmouth, but it did not prosper. At that time, Roy and his wife, Ethel, had a son, Eugene, born in 1928. Times were tough, and they were looking for a place to grow and sell vegetables. They settled in Hatchville and rented rooms. Their daughter Louise was born in 1931. In that year, they moved into the house of Lucina Swett, a small, simple farmhouse on Hatchville Road west of the Dimmocks. The house was the last to get electric lights in Hatchville and didn't have indoor toilets, according to Henry Jensen, who visited there. In exchange for caring for Mrs. Swett, they could live in her house. After Mrs. Swett died, the house and property (6.8 acres with a barn and a cranberry house) became theirs.

The Allens, of course, had a large family garden that saw them through the winter months, and for years, they ate what they could grow and raise, including chickens. In the winter, they got a cow and housed it in a barn, where it was fed and milked. They picked apples, which were put in a stone-lined cellar. On the property, according to Louise Allen Bailey, were at least twenty-five to thirty fruit trees, all single varieties.

Mr. Allen sold fruit and vegetables from his truck. Eugene Allen recalled helping his father on the truck when he was about eight or nine years old. Roy Allen also worked for Wilfrid Wheeler at Ashumet Farm for a short time. During the construction of Camp Edwards, Roy was a foreman of a crew building and laying out roads on a WPA-funded project. He later ran the Post Exchange at Camp Edwards until 1946–47. Later, Roy worked for the Falmouth Park Department until his retirement in 1960, when the Allens moved to Ransom Road in the village of Falmouth.

Both Gene and Louise recalled that there were few children to play with in Hatchville and that they felt somewhat isolated. There were the Baldic boys prior to 1939, Richard Dimmock, the three Williams children at Crooked Pond Farm and the Keelers on Geggatt Road. The Allen kids would swim in Coonamessett Pond in a relatively private spot. Apparently, the Dupees would allow them and Richard Dimmock to swim there but discouraged all others. In the winter, the children skated on the adjacent cranberry bogs.

Eugene's childhood chores included feeding the hens and milking their cow twice a day. He also had a *Standard Times* paper route (not on Sunday) extending from Hatchville Road to Boxberry Hill Road to Sandwich Road down to John Parker Road—i.e., most of historic Hatchville. Gene also worked for David Dimmock, helping with haying both on Hunky Dory Farm and on land leased by Mr. Jensen north of Coonamessett Road.

After graduating from Lawrence High School, Gene Allen worked at several jobs before entering the U.S. Army. The next year, at age twenty-three, he married and moved to North Falmouth. He worked briefly for WHOI before returning to the National Guard, where he spent thirty years and retired with good benefits.

Louise Allen attended Falmouth Public Schools and took a business course, graduating from Lawrence High School in 1949 in the same class as Greta Jensen. Louise soon found work at WHOI and with the town of Falmouth. She recalled that the family would attend church nearby at the East End Meetinghouse on Sandwich Road, which was open in the summer months.

The EEMH and its parsonage were often the site for Hatchville social events, such as regular potluck dinners. In 1945, Mrs. Dimmock, Mrs. Allen, Miss Mildred Hatch and Mrs. Randall organized a Halloween Party, sponsored by the program committee of the East End Congregational Society, for thirty-two young people at Pastor Thygeson's House, the former parsonage, which was used as a community center. Prizes were awarded to Henry Jensen for his Sherlock Holmes costume and to Fred Jensen for Magic Island (a science-based radio program). Refreshments were followed by music and square dancing.[94]

Dupee and Razinha Families at Lochstead Property

In contrast to the life of the nearby working farmers, the Dupee family came to historic Hatchville for its recreational and environmental amenities. Norman and Marjorie Dupee had a home in Waban, Massachusetts, but in 1937, they purchased Robert Leatherbee's 76.7-acre parcel that included the Lochstead manor house on the north shore of Coonamessett Pond to

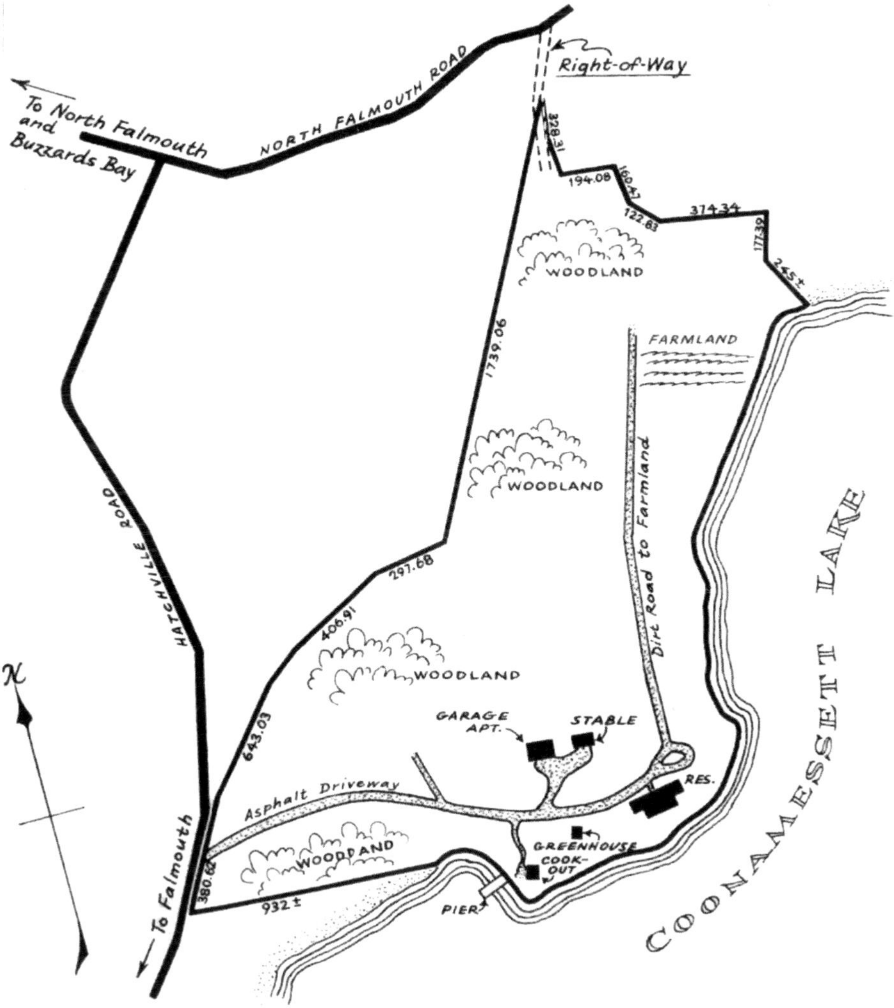

Lochstead property map as part of an offering brochure prior to its sale to G.L. Rowsey in 1956. *Courtesy Donna Jewett.*

use as a family compound during the summer. Norman and Marjorie had four children: Norman E. Jr. (aka Ned), Donald, Paul and Jane. Norman's adopted sister, Maya, ran the manor house and looked after the needs of the children and, later, the grandchildren. Norman hunted and fished. Like the sportsmen of an earlier generation, he found Hatchville a perfect locale.

In 1938, Lochstead had forty fowl, a stable, a manor house valued at $20,000 and a garage, above which was an apartment where permanent

Marjorie Dupee and James Razinha collaborated to grow roses at Lochstead for her to paint. *Courtesy Donna Jewett.*

caretaker James Razinha and his family lived. During the next twenty years, the history of this property is about the shared and separate events in the lives of the Dupees and the Razinhas.

James and Alice Perry Razinha and toddler Jimmy, born in 1935, came from the Whitney Estate in Woods Hole to work at the Lochstead property. Mr. Razinha took care of the landscaping, gardening and other outside work while Mrs. Razinha was responsible for cleaning the house and doing the laundry. The Razinhas received one salary with free housing and a car on a spectacular property.

Mr. Razinha and Marjorie Dupee grew ornamental plants and flowers, many propagated in a heated greenhouse. The flower garden was four elongated beds visible from the kitchen pantry and dining room. Marjorie's specialty was cultivating roses, which she then painted in classic settings in her studio. In the northeast corner of the property, Jim maintained a large vegetable garden as well as a peach and apple orchard.

Dupees walking toward the first tee of the resort's golf course. The Coonamessett Club, soon to be the Falmouth Playhouse, is in the background. *Courtesy Donna Jewett.*

Norman Dupee (standing) with family members at Lochstead boat dock on Coonamessett Pond. *Courtesy Donna Jewett.*

For most of the year, the Razinhas were able to live off the land, and everyone at Lochstead ate fresh vegetables and fruit during the summer. Carol was the Razinhas' second child. When she was older, her chores included weeding carrots and painting boats that her father built, which were rented to occupants of the cottages at the other Dupee property on Coonamessett Pond at Atamannsit.

The Hatchville Post Office on Boxberry Hill Road was a favorite place for local children. There were about twenty post boxes. Postmistress Mildred Hatch, who also taught at Mullen Hall School in Falmouth, made hand puppets, and when the children went to get the mail, she would put on a show for them. Carol Baker remembered that as a child she would walk through the Lochstead vegetable garden to the post office to see Mildred's black rag doll that all the kids coveted.

As the Dupee's children married and had their own children, they all would spend summers at Lochstead. The Razinha children were included in many Dupee family activities. Jimmy Razinha soon preferred the company of his friend Henry Jensen from the nearby dairy farm. Typical activities for the Dupee grandchildren included blueberry picking on the property, selecting food from the vegetable garden and putting on skits and plays. Dupee family gatherings inevitably centered on the adjacent Coonamessett Pond beach for clambakes and water sports. The Dupees golfed at the Coonamessett Resort and fished the nearby ocean waters in their substantial motorboats.

Carol and Donna Jewett, a daughter of Donald and Helen Dupee, were the same age. As one would expect, they became close friends and played together. Carol had her own small motorboat that they used on Coonamessett Pond, often to look for golf balls near the ninth hole of the golf course or trying to spot celebrities near the Star Cottage (a small abode away from the inn) and the Falmouth Playhouse. Today, these two friends can see each other's homes across Coonamessett Pond.

DUPEE FAMILY AT ATAMANNSIT

In 1938, Norman Dupee purchased the 150-acre Atamannsit property along the southwest shore of Coonamessett Pond, previously owned by George St. Amant, from the Roxbury Institution for Savings at foreclosure. The property included a bungalow, a hunting camp and the barns for St. Amant's prized Guernseys, which Dupee removed.

Norman Dupee subdivided the Atamannsit property among three of his four children. Cottages were constructed for summer rentals, which his son Ned managed. As the visiting grandchildren became too numerous (eventually there were thirteen) at the manor house on the Lochstead property in the 1950s, each of these adult children built summer cottages on the pond at Atamannsit. Eventually, at the urging of Helen Dupee, Coonamessett Lake was declared a "Great Pond" on which boat engines of no greater than six horsepower were permitted. The houses and cottages at Atamannsit eventually passed to the next generation.

LaBonte Land Transfer

In the 1960s, a single land transfer between parents and a married daughter would prove to be critical for the eventual establishment of two progressive agricultural institutions on Hatchville Road: New Alchemy Institute (NAI) and Coonamessett Farm. As described in Chapter 5, Alma LaBonte had acquired land from Robert Leatherbee's estate. By 1947, she owned eighty-seven acres along Hatchville Road on which were a lovely barn, courtesy of bankrupt Kensington Farm, and a house. In 1964, Alma and William LaBonte transferred their land to their daughter, Anita, and her husband, A. Robert Gunning, and moved to Florida.

Anita LaBonte grew up in Hatchville across the road from the Williams' Crooked Pond Farm. Sadly, her brother, William, had died of bronchial pneumonia in October 1937, a few months after Mrs. LaBonte had acquired the land on Hatchville Road. In 1939, Anita graduated from Lawrence High School, where for three years she was the only female golfer in the Intercollegiate Golf Association of Eastern Massachusetts. Taught the sport by her father, William, she regularly played golf at the Coonamessett Resort. William was a caretaker there, so Anita had no difficulty getting tee times.

Anita went to Cape Cod Secretarial School in Hyannis and then worked in the biology department at WHOI, from which she retired. In 1941, she married A. Robert Gunning, who was exempt from military service. He continued working at the Marine Biological Laboratory as building superintendent and offered his expertise to help maintain the structural integrity of the poorly maintained East End Meetinghouse. The Gunnings lived on Hatchville Road and had one son.

COONAMESSETT FARM (1989–PRESENT)

The Gunnings sold a sizeable parcel on the east side of Hatchville Road to Maurice Strong in 1979, which Ronald and Dr. Roxanna Smolowitz purchased from him in 1986. On that property, the Smolowitzes established Coonamessett Farm, a twenty-acre member-supported destination farm that they have operated since 1989.[95]

Coonamessett Farm entrance sign. *Photo by author.*

NEW ALCHEMY INSTITUTE (1972–1991)

On twelve acres of land leased from the Gunnings on Hatchville Road (just south of Coonamessett Farm and opposite Geggatt Road), Nancy Jack Todd, John Todd and William McLarney established the New Alchemy Institute (NAI), a nonprofit organization. Their mission was to experimentally determine and design how humans might live a sustainable existence using appropriate technology—a tall order, but one with great promise. As might be expected, the NAI research program included food, energy and shelter. Its innovative program attracted many followers and made historic Hatchville a destination for the new ecologists and back-to-the-land types.

NAI soon experimented with aquaculture tanks, a wood stove–heated geodesic dome greenhouse that contained a fishpond, an organic vegetable garden irrigated by fishpond water and a windmill. Fish (*Tilapia*) were harvested, cooked, eaten and even written about in the *New York Times*. A few years later, NAI completed the famous ark, a large solar greenhouse (a bioshelter) that housed fishponds, food crops, soil life and insect communities.

Gary Hirshberg and Greg Watson joined NAI in 1980. Gary stayed at NAI for about two years as executive director before moving to New Hampshire, where today he is chairman and former president and CEO of Stonyfield Farm. Greg Watson was the NAI education director until

New Alchemy Institute brochure cover depicting geodesic dome, windmill and ark. *Author collection.*

A 2008 photograph of the successor to New Alchemy Institute's free standing Ark (a bioshelter), which is now connected to a house and contains solar aquaculture ponds, plants and living space. *Photo by author.*

he, too, moved on in 1983, only to return in 1989. Greg is an alternative energy expert with many years in state government, who was appointed as Massachusetts's commissioner of agriculture in 2012, a fitting bookend to the first commissioner, Wilfrid Wheeler.

To provide some predictability for its research and education center, in 1982, NAI bought the land it had been renting at a cost of $300,000. Also that year, it built a super-insulated auditorium and set up a small farm stand to sell garden produce. However, for the next few years, funding for the institute's work was difficult to find. Still, reports were published on the economics of solar greenhouses and on organic farming in general. In 1986, the rear of the barn was renovated for a new visitors' center, and produce sales reached $10,000. The NAI tour in 1988 would have included the theme gardens with three hundred annual and perennial shrubs, berries, flowers, herbs and mounds of steaming compost.

At the end of 1990, NAI was facing a severe cash crunch. It appeared that the limited pool of funds available for sustainable agriculture research was being divided too many ways and that according to founder Nancy Jack Todd, "New Alchemy's work was no longer seen as indispensable."[96]

The NAI closed in 1991, and the land was sold to Cape Co-Housing, an environmental community of ten member families organized by Earle Barnhart and Hilde Maingay, who today maintain the Green Center.[97]

Publicity about NAI's provocative and attractive ideas for sustainable agriculture as well as its implementation of those ideas had attracted people from around the world to Hatchville to listen and learn. No longer was this part of the village just known for the sprawling Coonamessett Resort. New Alchemy's legacy is guaranteed, especially in Falmouth, where dedicated ecologists and organic gardeners actively seek solutions for both local and global environmental problems.[98]

PART III

CHALLENGES TO THE CHARACTER OF HISTORIC HATCHVILLE

Chapter 9
SALE OF COONAMESSETT RANCH COMPANY ASSETS

Frances Anita Crane's death in August 1954 led to a reassessment of the Cranes' future involvement with the Coonamessett Resort. As a result, between 1956 and 1958, the trustees of her estate and her equal partner, John O. Crane, sold the resort's assets. There were three separate sales: Falmouth Playhouse, Coonamessett Airport and contiguous land to the west and, finally, all other parts of the resort, including the golf course and Treadway Inn and Lodge. These sales gave the remaining Crane family (Frances's sons, Robert and Richard, and their uncle John O. Crane) an acceptable but belated exit from thirty-three years in the hospitality business in historic Hatchville. The consequences for the direction of land use in historic Hatchville of the sale of the resort, specifically, would appear in about twenty years.

After these sales, John Crane looked after some businesses' interests in Rome. During the war, his farm was occupied by the Germans and then bombed by the Allies, for which he sought some reparations for the damage. He later built a restaurant called *All'Acquedotto* across the road from his destroyed farm. The restaurant opened in 1968 and is there now.[99]

SALE OF FALMOUTH PLAYHOUSE

Sara "Sidney" Gordon purchased the Falmouth Playhouse in 1956 after four seasons as manager when Richard Aldrich withdrew as the nominal

Ruins of the Falmouth Playhouse, which burned in a suspicious fire on February 28, 1994. *Courtesy of Daniel Small.*

producer. Gordon was described as a "tall, vivacious red head," who would greet the playgoers personally on opening nights.[100] It appears that the star-studded performances booked by her were of high quality—headlined by Broadway and screen actors such as Helen Hayes, Ethel Barrymore, Sir Cedric Hardwick and Ann Southern—and financially successful. After sixteen years in charge, Sara Gordon sold the Falmouth Playhouse to Eddie Bracken Ventures in 1970 for $205,000, a deal that unfortunately ended in foreclosure a year later and cost her $100,000 to get the playhouse back.[101]

Gordon finally sold the Falmouth Playhouse to Ralph A. Miller in April 1984. Miller reportedly spent over $1 million dollars renovating the theater before it reopened in 1988. He didn't follow the same business model of stars acting in Broadway plays. In contrast, his company produced lighter theatrical fare, with an emphasis on dinner theater. The playhouse was a popular nightspot until the last day of February 1994, when it was destroyed by a fire deemed to be of suspicious origin.[102]

FRANCES A. CRANE WILDLIFE MANAGEMENT AREA

The Coonamessett Associates also wanted to sell the former agricultural land north and west of the Treadway Lodge, including the airport leased by George Pinto since 1955. At this time, the airport had three unpaved runways, with the longest, at 4,500 feet, to the northeast and strip lighting on the northeast–southwest runway only (common wind direction). All instrument flights were under control of the nearby Otis Air Force Base. Even though the Falmouth Aviation Committee had an interest in acquiring some or all of this property in 1957, it did not speak out in a timely manner, according to Pinto.[103]

Meanwhile, in 1958, the Division of Fish and Wildlife (DFW) purchased 1,562.1 acres of woodland and grassland north of Route 151, as well as the small airport, for $77,500. The land was to be used for wildlife management and named in memory of Frances Anita Crane. Unfortunately for Mr. Pinto, the airport would not be compatible with wildlife management practices. Pinto, nevertheless, argued that his lease with the Coonamessett Ranch Company ran for ten years, until 1965. The fight was on.

Pinto's supporters maintained that a year-round airport was more important to the town than a hunting preserve. Even two years after the sale, the Falmouth Town Meeting authorized the Falmouth Airport Commission to wrest the eighty-acre airport from the commonwealth (DFW), to raise approximately $20,000 and "to procure any legislation necessary in connection with such purpose." And if the commission couldn't buy the property, it was directed to lease it.[104]

The legislative route never got off the ground. Eviction was planned for September 10, 1960, but the saga continued. Finally, Executive Air Services, Pinto's company, was evicted on June 1, 1961. The state prevailed, and the Coonamessett airport was closed.

Gradually, the Frances A. Crane Wildlife Management Area expanded to its current 2,000 acres when the DFW acquired properties adjacent to the original purchase (the pheasant area) and later added 380 acres bounded by Currier and Hayway Roads and Route 151 (the quail area) about 1.5 miles east of Ranch Road.

The original purchase included approximately two hundred acres of abandoned farmland near the airport, likely the best cropland in Hatchville. The DFW found that on half the acreage, the topsoil had been removed to construct a runway. Also, the initial survey found that perennial grasses, weeds and various woody plants had invaded abandoned fields. Moreover, there was very little water for wildlife such as bobwhite, quail, cottontail rabbits, ruffed grouse, woodcock, pheasants, fox, raccoon, deer and gray squirrels. Wild dogs preyed on this wildlife.

The DFW executed a management plan in 1962 for the Crane Wildlife Management Area to provide small-game hunting with special emphasis toward native quail management and stocking of both quail and pheasants. The initial three years of the plan required a number of improvements: 1) planting fifty-five acres of agricultural crops, 2) replanting twenty-five acres of abandoned fields and gravel pits with perennial field mixtures, 3) planting nineteen hedgerows or field dividers, 4) clearing and closing fourteen miles of roadways and access trails 5) building five parking areas and 6) digging one well.[105]

The Frances A. Crane Wildlife Management Area (pheasant area), though primarily a place for hunting and dog training, today is used also for biking, nature study, bird watching and horseback riding originating from local stables. Also, somewhat fitting, the Otis Model Aircraft Club members fly colorful, sleek remote-controlled airplanes on one of the old runways they maintain.

CLAUSON'S RESORT (1957–1977)

Liquidation of the Coonamessett Ranch Company's resort assets culminated in June 1957 with the mortgage sale to Clauson's Garage Company, owned by Harvey G. Clauson Sr. Included were the golf course and the Treadway Inn, as well as the Treadway Lodge, and twenty-five guest cottages. The purchase also included hundreds of acres of Coonamessett Ranch Company land. According to public records, the note payable to the Coonamessett Associates was $219,083.35, although the *Falmouth Enterprise* reported the sale as $350,000.00.

George Clauson recalled the initial visit by Harvey G. Clauson Sr. to the Treadway Resort: "I remember coming to Hatchville with my grandfather. Sandwich Road had nothing on it. It seemed like we drove forever. We turned off on to Boxberry Hill Road, I asked, 'Where are we going?' 'I'm looking at something I want to buy,' he said. And I replied, 'Way out here?'"[106]

Harvey G. Clauson Sr. and his son Donald H. Clauson assumed the management of the resort from the Treadway Corporation. Managing a golf course suited the elder Clauson, an avid golfer who, in 1956, had owned the Blue Hill Country Club, the site of the PGA Tournament. But neither

Harvey G. Clauson Jr. with a friend at Clauson's Inn after the purchase of the resort by his father, Harvey G. Clauson Sr. *Courtesy of Harvey G. Clauson Jr.*

he nor his son knew how to run the food, beverage or other parts of the business, which was a very costly lesson. Tragically, in May 1958, Harvey G. Clauson Sr., a decorated World War I veteran, died in a house fire at his modest East Falmouth home; he was fifty-nine years old. He left his wife, M. Elizabeth Clauson, as his primary heir.[107]

Arnold Palmer (fifth from left) and Billy Casper (third from left) and others with Harvey G. Clauson Jr. (sixth from left) at a golf exhibition on August 13, 1960, held at Clauson's Inn and Country Club. *Courtesy of Harvey G. Clauson Jr.*

After his father's death, Harvey G. Clauson Jr., thirty-three, sold Clauson's Garage Company in Brookline, Massachusetts, which he had purchased from his father (the name was retained for various business transactions). He then moved to Falmouth to help his family run the resort. Although Donald Clauson was in charge of the inn and other resort facilities, it appears that his mother had the final say in most matters.

For a while, Harvey G. Clauson Jr. operated the golf course, where he excelled at attracting talented amateur and professional golfers and promoting the resort among professional athletes and their fans. Harvey Clauson's biggest coup occurred on August 13, 1960, when Arnold Palmer, Billy Casper and some regional talent played an exhibition match before a crowd of two thousand. On that very warm day, Palmer's team beat Casper's team in an eighteen-hole best-ball competition, in which the team member with the lowest score posts the team's score for that hole.[108]

For the Clausons, it should have been the best of times. Postwar prosperity finally was putting more disposable income into the pockets of the middle class, who traveled by family car and vacationed at full-service resorts. Cape

Cod's population had nearly doubled, which also increased regional interest in the resort. Moreover, the Cape was an affordable, attractive destination, quaint but not exactly ritzy, and the inn was a cut above the rest. At that time, Clauson's Resort offered four lodging options on the European plan. Prices for two persons per night were fourteen to eighteen dollars with a private bath at the inn, eighteen dollars at the Motor Lodge, twenty-four dollars at the Ranch and sixteen to twenty-eight dollars for the cottages. At the resort's restaurants, music was provided at cocktail and dinner times. There was even a water-ski school that held a Sunday afternoon exhibition.

Another advantage for the resort was that the fabulous Falmouth Playhouse was adjacent to its golf course. For guests, it was a short walk along the first fairway or a ride in an electric golf cart, which they could use to zip around the entire resort. Maybe unspoken but nevertheless important was that celebrities played golf and rubbed shoulders with the guests. The celebrity list, recalled by George Clauson, included performers Johnnie Mathis, Jerry Lewis, Mickey Rooney, Julie Newmar, Jose Feliciano, ZaZa Gabor, Lloyd Bridges and William Shatner, as well as professional athletes Wayne Embry, Bill Russell, Don Nelson and Tony Conigliaro.

Donald Clauson, like the Treadway group before him, also wanted to attract businesses and professional associations for sales meetings, retreats and conferences, so the resort ramped up its banquet services. Clauson's was self-contained and only ninety minutes from Boston, and thus, it would have been attractive to event planners. To accommodate large numbers, the resort had three different banquet rooms in the enlarged inn: the Terrace Room (seventy-five people), the Club Room (forty people) and the Fairway Room (one hundred people, where the buffet was served). These same spaces, plus the Lounge Room (one hundred people), the Studio Room (sixty-five people) and the Cranberry Scoop (twenty-five people), were available for meetings. Since it was open all year, Clauson's resort could also host many weddings.

GERMAN "JERRY" LOPEZ'S MISTAKE

Donald Clauson really needed an inn and food service manager like the talented Edna Harris, who had leased these services until 1953 (see Chapter 7). Mr. Clauson hired German "Jerry" Lopez to run the food and entertainment side of the business. Unfortunately, Mr. Lopez made a serious error of judgment when on October 24, 1961, he bribed an IRS agent by the name of Roger S. Davis who was investigating why Clauson's hadn't paid the excise tax on the apparent cabaret on the premises (there was dancing in the bar and lounge). Lopez suggested to Davis that he drop the case and offered the agent money. Lopez later testified in his trial that the cash was payment of the tax and for Davis to prepare his tax returns and help get his books in order.

Agent Davis reported the $420 bribe to his superiors and returned to meet Mr. Lopez wearing a recording device to collect more evidence but without a search warrant. Lopez apparently gave the agent more money. The recording was instrumental in convicting Lopez on three of four counts of attempted bribery of an Internal Revenue Agent for which he was sentenced to one year in jail. Presumably, Mr. Clauson replaced Mr. Lopez with another manager.

Lopez's attorney appealed the conviction. He objected to the admission in evidence of the recording of the agent's conversation with the petitioner as a violation of the Fourth Amendment. The appeal reached the U.S. Supreme Court in *Lopez v. United States*, 373 U.S. 427 (1963), which affirmed the decision of the court of appeal.[109] This case became a landmark federal decision regarding cases in which one party tapes another without permission, so-called one-party taping.

GOLF COURSE

Owning a golf course in the 1960s should have been highly profitable. All the factors pointed to success: a more prosperous middle class, more leisure time and a surge of new player interest.

After the initial two years when he devoted more of his time to managing the golf course, Harvey Clauson moved on to consolidate his automobile dealership and real estate businesses in Falmouth and left the running of the golf course to others.[110] David Mauk, a Hatchville resident, was an early hire and proved to be a trusted and knowledgeable golf course superintendent for forty years. In 1971, George Clauson, Harvey's son, became head professional, and he and Mauk managed the course for the next six years. Other professionals at the resort before George included Charles Borden, Ross Coon and Huston "Paddy" Leclair.

Predictably, as golf became more popular, competition for time on local courses like Clauson's became intense. According to George Clauson, his course was often at capacity, hosting about 250 golfers per day. As a result of this congestion and demand, several local golf courses were constructed: the Falcon Course on the MMR, just a few hundred yards from the closed Coonamessett airport, catered to military personnel; the Falmouth Country Club, founded and built by Louis Rabesa, Ernest Cabral, James Razinha and Anthony Cunha and now owned by the Town of Falmouth, opened in 1965; about the same time, New Seabury in Mashpee was built; and in 1968, PGA pro Paul Harney opened an eighteen-hole "executive course" less than a mile away from Clauson's on the north side of Route 151. Clauson's had a major advantage, however: golfers could stay overnight at its various lodgings and use the other resort facilities. Besides, its green fees were reasonable—five dollars on weekends and four dollars at other times.

NICKELODEON CINEMA

Movie theater developers Joel and Henrietta Tranum and their partners Murro E. and Joanne Van Neter opened this small art house in 1972 on three acres immediately to the west of the Clauson's Motor Lodge on Nathan Ellis Highway. Originally, it had two screens but expanded to five within a decade. The Nickelodeon was popular because the Tranums showed foreign, classic and independent films not usually distributed on Cape Cod.

According to Falmouth assessors' records, the Interstate Corporation (Barnstable Realty Company) purchased the Nickelodeon from the Tranums in 1986 for $1.1 million. Subsequently, Hoyts Cinema Corporation operated the Nickelodeon but closed it in 1990 for five years. In 2003, Regal Entertainment Group assumed control of the cinema until it was permanently shut in October 2012, presumably because of the lack of paying customers and to avoid of the expense of upgrading to new digital technology. In 2013, the cinema was sold to Generation AJW LLC for a reported $545,000. Generation plans to remove the theaters and convert the existing structure for use as a commercial landscape garage with two overhead apartments for its tenant, Hamilton Tree & Landscape LLC. This property at 742 Nathan Ellis Highway (Route 151) abuts the Frances A. Crane Wildlife Management Area to its north.

M. ELIZABETH CLAUSON SELLS RESORT

M. Elizabeth Clauson (Clauson's Garage Company) sold the resort, including the golf course, to Steps Realty Trust (Stephen Hopkins and Shepard Sumberg) on May 24, 1977, for $1.475 million. Mr. Sumberg, from Newton, Massachusetts, had developed homes in nearby Ashumet Valley and was an investor in the Sheraton Inn-Falmouth. He had been the New England broker for the Lake Havasu, Arizona/London Bridge development. Mr. Hopkins was a land developer and mortgage broker in Woburn and Newton, Massachusetts. For their mortgaged investment, the pair received 375 acres of land, the resort buildings, and the golf course. However, Clauson's Garage Company retained about sixty-five acres of resort land, which would become a valuable asset for Harvey G. Clauson Jr.

Before too long, it was obvious to the Hatchville community that Steps Realty Trust was not going to operate a resort. Now, with the change in ownership of the resort, uncertainty reigned: would it become open season for subdivision proposals that included the golf course and surrounding undeveloped, unprotected land?

Chapter 10

HATCHVILLE DEVELOPMENT SCHEMES

Developing residential home sites in the late 1970s and early 1980s created both the opportunity for suburban living on Cape Cod and the demand for it. Even though the Cape's infrastructure, especially transportation, was still mediocre at best, people came to stay. Many home purchasers had reached retirement age, while others obviously decided that commuting to jobs off-Cape by crossing the Bourne or Sagamore Bridges over the Cape Cod Canal was worth the hassle. The result was a near doubling of the population between 1960 and 1980 in both Barnstable County (147,925 in 1980) and the town of Falmouth (23,640 in 1980). Few brakes were applied to this rampant development. Agriculture and other open space, except for golf courses, too often were an afterthought. In the process, much of what made Cape Cod attractive to these newcomers and precious to others was being lost. Still, more people came. In response to these changes, the Association to Preserve Cape Cod (APCC), which was formed in 1968, advocated for protection of water resources, land preservation and smart growth.[111]

Specifically in Falmouth, 3,754 housing units were built in the 1980s, and lightly developed Hatchville became a rich target for expanded residential development.

DEVELOPMENT OF WESTERN HATCHVILLE

On the western edge of historic Hatchville, the first residential developments appeared slowly during the 1970s and early 1980s, some of it on land once owned by Peter Gomes. Houses were built on a new subdivision north of Deep Pond between Boxberry Hill and Sam Turner Roads, where Goeletta Drive is the longest road. Typically, these were small houses on small lots. Houses also went up west of Sam Turner Road. These semirural developments were attractively priced and adjacent to the promise of relatively rapid automobile travel on the Nathan Ellis Highway (Route 151) and Route 28 to the Bourne Bridge.

Further expansion westward toward North Falmouth was more challenging. The northwest corner of present-day Hatchville is mainly rocky glacial outwash east of the Buzzards Bay Moraine, with hills and valleys covered with scrub oak and pine and many, many boulders. Not until the 1980s did anyone consider developing this area for commercial or residential use, and even then, the first steps were tentative: plans for new roads on 133 acres were still not finalized by 1986.

It took another three years before construction began in western Hatchville. In 1989, a country club and an eighteen-hole golf course called Ballymeade opened in the Falmouth Woods development. The country club, the developers offered, would be the centerpiece of a 258-single-family-home community. However, before the residential lots could be sold, in 1990, the development's promoters, Patricia and Stephen Harrison, went bankrupt. The next year, they were indicted on one hundred counts of bank fraud and other charges. The unsealed indictment released by the U.S. Attorney's Office on December 12, 1991, alleged that from 1985 to 1989 the Harrisons defrauded both Sentry Federal Savings Bank of Hyannis and the Bank of New England, Boston. Sentry had loaned them $10 million and Bank of New England $28 million.[112]

Ballymeade Development Corporation (CLSV Associates Limited Partnership, John T. Callahan III, trustee) picked up the pieces and, in 1994, began to sell lots and homes after purchasing the Falmouth Woods development property at auction in 1992, reputedly for $6.2 million. Today, the well-kept roads, golf course, clubhouse and fitness center, walking trails and adjacent conservation land make the 250-plus home sites on about five hundred acres an appealing residential community.

In 2002, the Town of Falmouth bought for $5.6 million from the Ballymeade developers two parcels, one of 115 acres and another of 14.6

acres, to add to the moraine green corridor. A private golf course, "The Golf Club of Cape Cod," opened on the east side of the Ballymeade property in June 2007.

DEVELOPABLE LAND IN HISTORIC HATCHVILLE

Looking back to 1977, the last year of Clauson's Resort, the most attractive land for large-scale residential development in historic Hatchville was within a polygon bounded by Route 151, then south on Sandwich Road, west onto Hatchville Road to Sam Turner Road and north to Route 151. This polygon contains Coonamessett, Crooked and Deep Ponds. The largest undeveloped parcels north of Coonamessett Pond were 1) land owned by the Jensens between Boxberry Hill Road and Route 151, 2) the 76.8-acre Lochstead property then owned by Alfred Morse and 3) the undeveloped parcels that were part of Clauson's Resort adjacent to the golf course north of Boxberry Hill Road and those parcels west of Ranch Road, including the golf driving range, adjacent to Route 151. South of Coonamessett Pond, the Atamannsit property, though divided among the Dupee family members, was also ripe for development.

Not surprisingly, Harvey G. Clauson Jr. would find business opportunities within the northern part of historic Hatchville. Mr. Clauson seemed to have two major passions in his life, golf and making deals that would bring him not only financial rewards but also respect. And make no mistake about it: he was a good golfer and businessman. He was smart and aggressive, rather a potent mix. It appears that Mr. Clauson made as many enemies as friends. For example, he was eulogized in 2007 as "a force to be reckoned with for sure, Harvey lived his life doing it his way."

In the southern part of historic Hatchville, Norman and Marjorie Dupee's grandson, Paul R. Dupee Jr., a self-described private investor, would find business opportunities within his family's Atamannsit property.

EFFECT OF THE SALE OF CLAUSON'S RESORT

Hundreds of acres of developable land within the aforementioned polygon were still in play after Steps Realty Trust purchased Clauson's Inn and

Resort in 1977. To partially finance the $1.475 million purchase, Steps Realty Trust sold the golf course to John Gilgun, former mayor of Woburn, Massachusetts, and golf professional Ernest Doherty. The mortgage was for $350,000. Gilgun and Doherty renamed the course—"Cape Cod Country Club"—and later built a condominium on the sixteenth hole (current address is 290 Boxberry Hill Road). Other financing for the resort purchase came from two banks and Clauson's Garage Company.

The buyers' intentions were unmistakable: Hopkins and Sumberg of Steps Realty Trust proposed that the former Clauson's Inn would be the hub of a 250 single-family-home development (each priced around $40,000) adjacent to the newly named golf course and along Ranch Road. Preliminary plans put the development costs at $5 million.[113] All the pair managed to develop, however, were a few residential home sites on Boxberry Hill Road and along Ranch Road. According to Ranch Road resident Louise Houle, Hopkins and Sumberg promised a freshwater supply, refurbished tennis courts (there were only two remaining from the resort at the time of the sale) and private beach rights on Coonamessett Pond for those home sites.

By the spring of 1980, Steps Realty Trust had done very little to maintain the resort's assets or develop additional home sites. Unfortunately, because of neglected roof repairs, most of the core resort buildings, including the inn, suffered water damage and much valuable wood was lost. Out of money, Hopkins and Sumberg were soon out of the picture. A legal judgment against them led to foreclosure proceedings, and Steps Realty Trust entered into Chapter 11 dissolution of its assets.

That Steps Realty Trust couldn't fund its planned residential development might have been good news to those concerned about preserving the rural character of the village. On the other hand, separating the resort property and its assets from the 148-acre golf course run by other parties (Gilgun and Doherty) made residential development more rather than less likely. The future of this part of historic Hatchville would still be precarious.

OPPORTUNITIES FOR HARVEY G. CLAUSON JR.

Harvey G. Clauson Jr.—golfer, car salesman and real estate developer—always hovered around the Clauson family's Hatchville business affairs. It wasn't long after the sale of the resort that he gained control of the Clausons' position as a creditor of Steps Realty Trust. How he did it, we probably will never know.

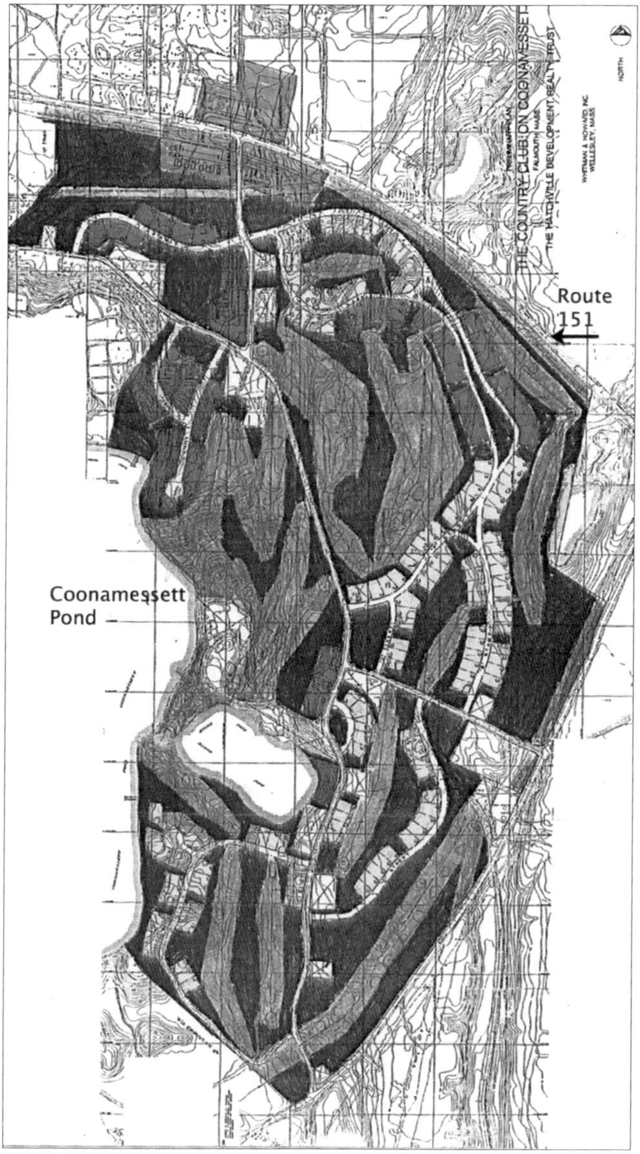

Plan of the 336-acre "Country Club on Coonamessett" proposed in June 1985 by Hatchville Development Realty Trust. Note the reorganized golf course with home sites. *Courtesy of Falmouth Planning Department, annotations by author.*

But his first action was to change the name of the creditor from Clauson's Garage Company to Boxberry Hill Land Company, of which he was the sole trustee.

The Bankruptcy Court authorized mortgage repayments to Suffolk Franklin Savings Bank ($63,200) and Cape Cod Bank and Trust ($700,000) and a note repayment to Boxberry Hill Land Company of $310,000 with interest.[114]

In addition, the court granted Boxberry Hill Land Company authorization to sell 216 acres of former Steps Realty Trust surveyed housing lots along Boxberry Hill and Ranch Roads to Silent Ridge Trust for $1,518,000. Mr. Richard Snider was the trustee of Silent Ridge Trust. The appearance of a new party, Silent Ridge Trust, meant that some of historic Hatchville land was likely to be developed.

Under court authorization, Phipps Products Corporation, a Massachusetts Chemical Company owned by Aaron Rosenberg, an associate of Mr. Snider, would guarantee the first two payments of $164,000 to Boxberry Hill Land Company.

On November 19, 1981, Silent Ridge Trust transferred the Hatchville real estate to Phipps Products Corporation. And on December 31, 1981, Phipps Products transferred the Hatchville land to another of Rosenberg's companies, Oliver Street Corporation.

Perhaps Oliver Street Corporation planned to hold on to this historic Hatchville land and catch the rising property values in this runaway development boom in Falmouth. Patient it was, because in four years its land along Boxberry Hill and Ranch Roads would become part of a mammoth planned subdivision, and they would make lots of money.

HATCHVILLE DEVELOPMENT REALTY TRUST PROPOSES LARGE SUBDIVISION

On July 1, 1985, Ernest Doherty and John Gilgun sold "Cape Cod Country Club" to Phil Friel of Green Meadow Golf Club Inc., reportedly for $1.625 million.[115] At the time of the sale, Cape Cod Country Club was already an important component of the preliminary plans for a 336-acre subdivision called "The Country Club on Coonamessett," which was filed with the Falmouth Planning Board on June 12, 1985, by Hatchville Development Realty Trust (HDRT).

As a partner, Oliver Street Corporation would supply to this planned development the surveyed and other lots it owned adjacent to the golf course. HDRT planned many single- and multifamily dwellings clustered around the reorganized golf course on four miles of new roads, as well as a twenty-two-acre commercial zone on both sides of Route 151 at Ranch Road.

Alert to the consequences of rampant unplanned development, two new groups emerged. In February 1985, several months before HDRT's preliminary plan reached the Falmouth Planning Board, a conference on the Cape Cod "land rush" inspired the notion to preserve three hundred acres of open space, one for each year since Falmouth's founding in 1686. Some of the attendees began to screen for possible purchase of undeveloped parcels that had ecological, recreational and social value. Appropriately, they named their organization the "300 Committee" and sent out two hundred letters to Falmouth residents explaining the need to acquire open space. Positive replies encouraged the group. Soon, the fledgling organization incorporated as a private, nonprofit land trust working to preserve and protect natural lands in Falmouth. The membership quickly grew to more than 1,500 families.[116] In direct response to the plans for "The Country Club on Coonamessett" in their immediate neighborhood, some Hatchville citizens formed the Coonamessett Pond Association (CPA).

Protracted negotiations between the Falmouth Planning Board and HDRT continued until February 4, 1986, when by a vote of four to two the latest plan was approved, with twenty-nine conditions. "The Country Club on Coonamessett" included a cluster plan of 225 units on 359 acres. Its plan also included twenty-two acres of commercial development along Route 151 and a proposal to donate four acres along Sandwich Road to the Portuguese American Association.

COONAMESSETT RESERVATION BORN

After the Planning Board's February vote to approve HDRT's subdivision plan, The 300 Committee began an elaborate campaign to persuade the citizens of Falmouth that it was the right time to preserve valuable open space. The committee chose ten parcels and sought purchase and sales agreements for them prior to the April special town meeting. The events at the town meeting of April 7 were unprecedented.

Mr. Bruce Tripp, chairman of the Conservation Commission, presented the warrant article to the town meeting, emphasizing the recreational, ecological and social value of each parcel. By a vote of 175 to 18, the town acquired eight tracts of land totaling 401 acres for $8.3 million, 100 acres more than they had originally sought to preserve for the town's tricentennial.

Within the "Coonamessett Pond, Hatchville" tract of 212 acres, the largest and most expensive of these eight parcels, was the new well site on Coonamessett Pond, which became operative in 1989 to replace an MMR-

Monument at the entrance to the Coonamessett Reservation at the intersection of Ranch Road and Route 151. *Photo by author.*

contaminated well farther north. The "Coonamessett Pond, Hatchville" parcel cost $3.9 million. Hatchville Development Realty Trust received $850,000. Its trustees were Richard B. Jones, Josiah K. Lilly IV, J. Theodore Dailey and William Mattos. Oliver Street Corporation received $3,050,000. A week later, a townwide referendum to authorize the borrowing for the land acquisition garnered over 70 percent of the vote. Thus the Coonamessett Reservation was born.[117]

As part of the purchase agreement, in May 1986, the Portuguese American Association received three contiguous lots at the junction of Ashumet and Sandwich Roads, on which it built the Falmouth Navigator, a banquet and meeting room facility.

Falmouth's Department of Public Works soon removed the derelict Coonamessett (Clauson's) Inn, which had stood for more than one hundred years overlooking Coonamessett Pond, first as the Chadwick's farmhouse and later as an expanded inn—so many memories gone.

Managing open space was a new experience for the Town of Falmouth. The 300 Committee and the Coonamessett Pond Association funded the Coonamessett Reservation Management plan, which was implemented in

1995. Much of the labor of grassland restoration and trail maintenance at the Coonamessett Reservation relies on volunteers, some of them local from the CPA and others from AmeriCorps.

1986: SECURE OR NOT?

In 1986, nine years after the sale of Clauson's Resort, secure open space in historic Hatchville included the Frances A. Crane Wildlife Management Area and the Coonamessett Reservation. The nearby Ashumet Holly and Wildlife Sanctuary was two decades old and adequately supported by the Massachusetts Audubon Society and a group of active volunteers. Still, there were properties around the edges of these open spaces that could be developed, especially the golf course, owned by Friel Golf Management.

Lochstead Association on Coonamessett Pond

While the Coonamessett Reservation was being created, Design Housing Inc. of Dedham, Massachusetts, Arnold Soloway's firm, was developing the 76.8-acre Lochstead parcel. This property was previously owned by Eugene Hatch, then Robert Leatherbee and later Norman and Marjorie Dupee. Norman Dupee died in 1956, and his widow, Marjorie Dupee, sold the Lochstead property to oilman G.L. Rowsey. The offering brochure listed the asking price as $85,000. The Razinhas moved to a new home they built on four acres on the east side of Coonamessett Pond on land willed to them by Norman Dupee.

G.L. Rowsey sold the Lochstead property in 1961 to shoe magnate Alfred Morse. In 1982, Frank Morse and Herbert Field created the Lochstead Trust to develop pond-side homes. And in 1986, the Lochstead Association was formed for a new residential development.

The Lochstead development had Falmouth Planning Board approval for seventy-three units.[118] Eventually, sixty-nine homes were built in a natural setting on Coonamessett Pond. The residents, who own their properties, also pay dues to the Lochstead Association for various amenities, including the restored 1925 manor house, which is the association's social center.

CHANGES AT ATAMANNSIT

The other Dupee land, the Atamannsit property and adjacent land on Coonamessett Pond with access to Sandwich Road, though not secure in 1986, was not much different than it was in the 1950s when Norman Dupee subdivided the property among three of his four children. Norman E. Dupee also had owned land at the junction of Hatchville and Sandwich Roads just north of the East End Meetinghouse. In 1986, the Falmouth Jewish Congregation (FJC) purchased this land from Norman E. Dupee Jr. and built a center for religious, educational and social functions. At the time of the sale, Norman E. Dupee Jr. was living in Texas. This parcel was valued at $325,000. The purchase price was $270,000, but Mr. Dupee gave a charitable gift of $55,000 to the grantee.[119] About 3.2 acres of that land was sold in 1992 by the FJC to the Unitarian Universalist Fellowship of Falmouth Inc.

In 1996, Paul R. Dupee Jr. purchased from his family members two houses and all the rental cottages at Atamannsit. He resold some of the dwellings but kept and renovated the lodge for his personal use. Eventually, this small, somewhat secluded residential community of fourteen dwellings (which were constructed either in 1920 by George St. Amant or in 1950 by the Dupees) that had a Hatchville Road address would be listed on private Atamannsit Road. Subsequently, only two new homes were built within this community, one in 2007 and another in 2011. Today, David and Donna Jewett (daughter of Donald and Helen Dupee) of the Dupee family live year round in Donna's parents' renovated house.

Paul Dupee Jr. also owned fourteen acres on Sandwich Road south of the Razinhas on Coonamessett Pond. On this parcel, Welch Healthcare of Norwell, Massachusetts, proposed an oversized retirement village. The project, which required a zoning change, did not achieve the two-thirds vote for approval at the Falmouth Town Meeting in the spring of 2002.

During 2002, as the Welch Healthcare plan was being resolved at the town meeting, Mr. Paul Dupee Jr., through his Falmouth attorney, negotiated with The 300 Committee for its purchase of most of the Atamannsit land he controlled. Also included was the fourteen-acre Sandwich Road property mentioned above, which he sold separately to the Town of Falmouth for $1 million of non-land bank funds, to be used as athletic fields. The largest part of this package included the Atamannsit woods and the meadows where St. Amant's expensive Guernseys once grazed, as well as fields and bogs that Norman Dupee had acquired. When the deal was completed, Mr. Paul

The Falmouth Jewish congregation has owned the old East End Meetinghouse since 1982. *Photo by author.*

Dupee Jr. sold sixty-two acres of conservation land, known as the "Dupee Parcel," for $4.1 million to the Town of Falmouth. The purchase was funded by the Land Bank, which is funded by a surcharge on property tax bills.

The Dupee Parcel was an important acquisition for Falmouth. It is adjacent to the twenty-five-acre Matt Souza Conservation Area, established in 1974, that follows the Coonamessett River south between Coonamessett Pond and Hatchville Road. Both the Souza and the Dupee lands, because they are adjacent to water, are critical vegetation buffers. Dr. Betsy Gladfelter of the Falmouth Conservation Commission explained their value: "These buffers not only reduce nitrogen and other kinds of pollutants and sediment from going into the water bodies but the roots of the trees and shrubs support a rich community of tiny invertebrates that actually break down materials as water migrates to the pond."[120]

OTHER HARVEY G. CLAUSON JR. HATCHVILLE SCHEMES

Harvey G. Clauson Jr. aggregated some parcels (recall the sixty-five acres not sold to Steps Realty Trust) within historic Hatchville that he sought to develop. In 1987, he and his son George Clauson, then of Clauson's Chevrolet and Oldsmobile, a Falmouth dealership, submitted plans to develop a new Chevrolet dealership at the southwest corner of Sandwich Road and Route 151. The dealership did not materialize, but Harvey Clauson still owned the parcels. Fifteen years later, in July 2002, the Falmouth Planning Board considered a proposal for a gas station and convenience store on that same corner, which Harvey Clauson called "Clausons Corner."[121] Before he sold the parcel for that same purpose, he placed a boulder on the corner inscribed: "Welcome to Hatchville 1683, Clausons Corner 1957." This boulder is the only place, other than Hatchville Road signs, which tells passersby that they are in the village of Hatchville. Immediately south of Clausons Corner on Sandwich Road, Alfred Clauson, Harvey's other son, developed five residential homes very near Little Jenkins Pond.

Harvey Clauson was involved in a bizarre series of events between 2001 and 2003 that caused consternation among the golf course neighbors. In 2001, Coonamessett Highlands LLC (established by sole trustee Harvey G. Clauson Jr.) purchased a one-acre parcel (Lot 162) northeast of the condominiums at 290 Boxberry Hill Road adjacent to the Cape Cod Country Club. On this lot, Clauson proposed a nine-unit residential development. Clauson's Garage Company had owned the lot before Mrs. Clauson sold the resort in 1977, but after that time, it was owned by John Gilgun and Ernest Doherty, who operated the Coonamessett Country Club golf course and who had built the condominium on the sixteenth hole. Gilgun and Doherty sold this one-acre lot to Robert F. Connolly in 1983 for $25,000. Connolly sold the lot to Coonamessett Highlands LLC for $60,000, and then things got very complicated. Gilgun and Doherty still owned the condominium, as mentioned above, but they had sold the golf course in 1985 to Phil Friel of New Cape Cod Country Club. Clauson's proposed development on this acre lot was opposed by both the Coonamessett Pond Association and the New Cape Cod Country Club, for similar reasons: the proposed development was too large, and it would force an unwelcome change in the golf course layout that could make it more rather than less likely that the golf course would be sold and or developed. To forestall the development, New Cape Cod Country Club Inc. sued Coonamessett Highlands LLC and the Falmouth

Marker at Clausons Corner, the intersection of Route 151 and Sandwich Road, the only public place (except for a road sign) where one sees the village name. *Courtesy of Maureen McIver.*

Board of Appeals. Eventually, through mediation, Coonamessett Highlands LLC sold the one-acre lot to New Cape Cod Country Club Inc. for $300,000 in May 2003.

This contentious event at the golf course and Mr. Clauson's aggressive development plans made the Coonamessett Pond Association and other Hatchville residents even more vigilant. The 148-acre golf course, which is a crucial piece of open space, could be sold at any time, and rumors circulated about its future. Those rumors had substance. An offering document from Grubb & Ellis Company of Newport Beach, California, in 2004 solicited bids for Cape Cod Country Club, touting many allowable land uses such as residential housing, schools, hospitals, retail and restaurants. The closing date for bids—April 1, 2006—passed without a change in ownership; meanwhile, significant improvements to the golf course have been made.[122]

During 2002, some members of Harvey Clauson's family and social circles received big payouts from the sale of historic Hatchville land to the Town of Falmouth, and presumably, others did not. Briefly, Harvey Clauson had his eye on a twenty-four-acre parcel (Lot 182, surveyed in 1992) west of the Coonamessett Reservation along Route 151 that was once part of Clauson's

Resort land. At one time, Harvey Clauson proposed a nine-hole golf course on this lot. Decades earlier, Robert Leatherbee of Brae Burn Farms had leased this parcel from the Coonamessett Ranch Company. To cement his position in Lot 182, in February 2002, Harvey bought the rights to the lot from three family members: Donald Clauson (brother), Arlene Robbins (sister) and Elizabeth J. Clauson (daughter); each was paid $80,000 to assign his or her right in Lot 182 to Harvey's son Alfred Clauson of Braeburn Farms Realty Trust.[123] As described below, it was an astute investment.

In July 2002, $985,000 of Land Bank funds were used to purchase properties Harvey G. Clauson Jr. had aggregated: to enlarge the existing Coonamessett Reservation, the Town of Falmouth acquired 23.89 acres (Lot 182 mentioned above) and 4.42 acres of woodland known as Turpentine Trail just east of Lot 182. The beneficiaries of this payout were the trustees of Brae Burn Farms Realty Trust, who were Alfred Clauson and Jeffrey White, a son of Harvey G. Clauson's companion Ellen White. And in order to protect Little Jenkins Pond near Clausons Corner, the Town of Falmouth acquired 3.86 acres around Little Jenkins Pond from One Fifty One Realty Trust (Jeffrey White, trustee).[124]

Volunteers, working with the Falmouth Conservation Commission, crafted walking trails through the woodland east of Ranch Road and west of Little Jenkins Pond.

Since those 2002 land transactions within historic Hatchville, The 300 Committee has championed an expansive "Coonamessett River Greenway," culminating in two critical purchases that should help preserve the continuity of this portion of the river. In 2007, the 10.0-acre River Bend Conservation Area, which contains the ceramic silo, was purchased for $1.2 million of private funds, followed in 2009 by the acquisition of the 6.8-acre David and Carmela Bartolomei Conservation Area at a cost of $1 million, derived from private and public funding sources.[125]

After these attempts to alter it, historic Hatchville has not lost the feel of horse and farm country. It is fair to ask: What is its future?

Coda

FUTURE OF HISTORIC HATCHVILLE

The previous ten chapters have, I hope, demonstrated that historic Hatchville is a special place with a unique history and that this book is a useful record of how things came to be the way they are. Of course, the interactions among many people with one another and with land made it a special place. One of my goals was to delve into the lives of the most interesting, influential and steadfast people for whom I could obtain quality information and tell their story in context. Notwithstanding this effort, I probably omitted some people and missed the importance of some events. I hope others will seize on these omissions as an opportunity to add to the historic Hatchville narrative.

In retrospect, I see no reason to reject the hypothesis I proposed at the outset of this investigation that a number of improbable land acquisitions and other events during the last 125 years led to the oldest part of the current village of Hatchville becoming an inland, semirural island in a sea of tree-shrouded suburbia.

My list of improbable events starts with Charles R. Crane coming to Woods Hole in 1901 followed by the 1916 formation of the Coonamessett Ranch Company and its purchase of 2,500 acres of historic Hatchville land and farms, which led to a model farm and later a large resort. Two other events that could be considered somewhat unlikely were the founding of two large dairy farms in Hatchville: George St. Amant's Atamannsit Farm on the southwest side of Coonamessett Pond, which began as a hunting camp, and Robert Leatherbee's Brae Burn Farms on the northwest edge of Coonamessett Pond. That these farm properties as well as most of the resort land remained largely intact into the twenty-first century is, I feel, also rather improbable. As a result, within historic Hatchville there is still a strong sense of horse and farm country, which is preserved because it is buffered by thousands of acres of open space and because it has some zoning protection.

I was reminded of this special sense of place by Rabbi Elias Lieberman of the Falmouth Jewish Congregation, who values living and working in historic Hatchville. He said that very few rabbis would be able to walk out the front door of their synagogues after a high holy day service and hear the sounds of horses coming from a stable across the road.[126]

What about the future? Thirty years ago, before the Coonamessett Reservation was founded from land that was once part of the Coonamessett Ranch Company's farm and resort, we could not have predicted current land use in historic Hatchville. As described in Chapter 10, without citizen action, it easily could have been hundreds of single-family homes extending from Route 151 to Coonamessett Pond and no Coonamessett Reservation. It is possible that ten years from now things might be different. Land use can change. Dr. Henry Dick of the Coonamessett Pond Association counseled that open space requires vigilant supporters: "There are a lot of pressures on land these days, and once land is acquired by the town doesn't mean it will stay as open space." He added: "One pressure is for municipal uses."[127]

How then should we be thinking about open space, including farmland? Certainly not as static. Currently accepted research informs us that Earth's climate is changing, and it does not seem likely that humans will be able to reverse these global effects to which we have contributed. What is not clear at this time is whether the pace of climate change is gradual or rapid.

Climate change, though measured globally, is being experienced locally. For example, historic Hatchville has thousands of acres of open space, much of it grassland; even golf courses and farms are open spaces. Questions arise: Will they undergo gradual or rapid ecological succession of plant and animals? Will new invasive species take over? Will there be new and different pests and diseases? Will these areas burn more readily? Will the water table rise or fall to extreme levels? These are all valid questions that can be answered if these lands are monitored. Such a monitoring process could result in proposing alternative futures—let's call it local adaptation—for each open space parcel, no matter how they were obtained.

At the seashore, change is occurring, too. Sea level is rising along the sixty-eight miles of Falmouth's coastline. People may elect to abandon the coast

Hatchville Day T-shirt design, depicting cranberries, herring, the meetinghouse, golf and military neighbors. *Courtesy of Danielle "Nellie" Kurz.*

and move inland, elevate their homes to anticipate severe flood conditions or just move away from Falmouth. Moving inland would put the sort of pressure on historic Hatchville and its inland open space that Henry Dick predicted could happen.

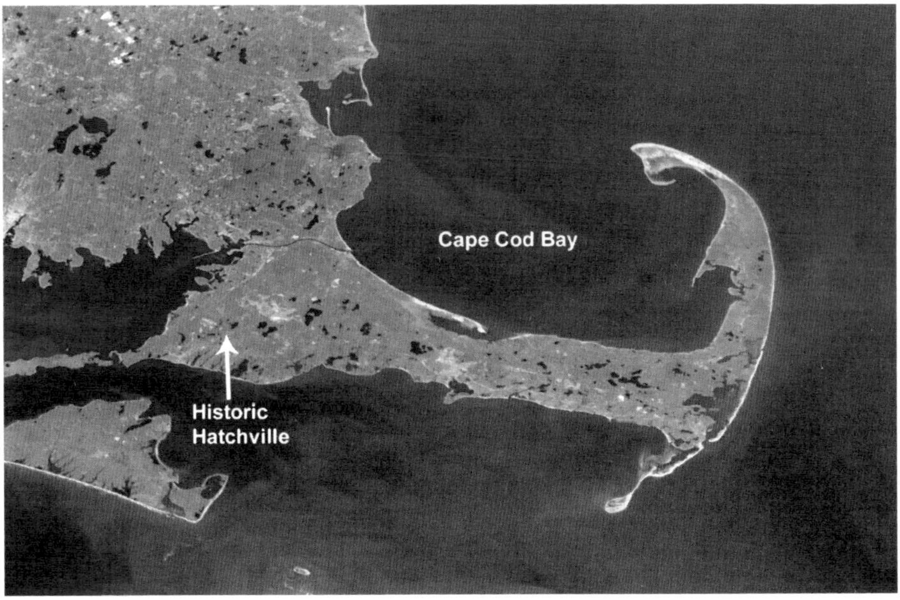

Satellite image of Cape Cod, showing the geographic location of historic Hatchville. *Public domain.*

Most people who care about open space in Falmouth seem content with the status quo. I suggest, however, that there should be much more time and energy invested to engage the public to use the existing open space not just in Hatchville but also throughout the town. Models of public participation that might stimulate interest are the August 2007 "Hatchville Day," at which over four hundred people took part in events at Coonamessett Farm and tours around historic Hatchville, and West Falmouth Days, which has been supported by 150 families in October 2012 and in 2013.

The more visitors to the town's open spaces, the more likely we will have informed decisions regarding the future of open space. As a corollary to this increased use, why not bring school-age children outdoors to these natural classrooms? They will be excited by the experience of discovery and want to return again and again. Such activity just might reverse the so-called nature-deficit disorder described by Richard Louv.[128] He maintains that direct exposure to nature is essential for the physical and emotional health of children and adults. I can't argue with that thesis—there is nothing to lose.

Ultimately we are left to answer the question: Who will determine the future of Hatchville? Hatchville resident Jay Zavala expressed his opinion: "Are there enough of us in Hatchville to address this [the future] from a political standpoint? Do we speak in unified voice about what we want for Hatchville or do we simply leave it to the planners, boards and commissions to tell us what we should be doing?"[129]

My hope is that historic Hatchville retains its essential character as a special, wild place within Falmouth and that it does not become a mere theme park or worse.

Appendix

HISTORIC HATCHVILLE TIMELINE

SEVENTEENTH CENTURY

1660: Jonathan Hatch and other Barnstable men establish a plantation on southwestern Cape Cod known as Succanessett.
1686: Town of Falmouth receives charter from Plymouth Colony.
1687: Road is built to gristmill on Five Mile River (also called Dexter's Road, Coonamessett Road).
1691: Falmouth is expanded to East End with purchase of land from the Wampanoag.

EIGHTEENTH CENTURY

1712: The 1691 so-called new purchase is laid out as hill and plain lots.
1740: Ebenezer Hatch (1708–1796) settles in the East End.
1797: East End Meetinghouse is built on 1.1 acres on Sandwich Road donated by Ezekiel Robinson.

NINETEENTH CENTURY

1821: East End Meetinghouse organizes as the Second Congregational Church of Falmouth.
1833: Silas Hatch is born in the East End.
1850: Falmouth population 2,621.
1855: Richard Teller Crane starts foundry business in Chicago, Illinois.
1858: Silas Hatch is appointed postmaster of Hatchville with the post office in his home.
1860: Census, Falmouth population 2,456; Hatches make up one-fourth of the 149 people enumerated in Hatchville.
1867: First one-room Hatchville school is built.
1870: Falmouth population fell to 2,237.
1880: Falmouth population 2,422.
1890: Falmouth population 2,567.
1891: Investment occurs in Hatchville cranberry bogs; investors include Joseph Story Fay; Azoreans buy land and farms in Hatchville.

TWENTIETH CENTURY

1900: Falmouth population 3,500.
1901: Charles R. Crane becomes a trustee of Marine Biological Laboratory in Woods Hole, Massachusetts.
1903: New one-room Hatchville school built.
1907: Robert W. Leatherbee marries Frances Anita Crane, daughter of Charles R. and Cornelia Crane of Chicago; Leatherbee works at Crane Company.
1910: Falmouth population 3,144; Charles R. Crane establishes family compound in Woods Hole.
1912: Richard Teller Crane dies in Chicago.
1914: Richard Teller Crane Jr. assumes control of Crane Company. Charles R. Crane retires from Crane Company and moves to New York City; Robert and Frances Anita Leatherbee relocate from Chicago to Woods Hole; agriculturalist Wilfrid Wheeler purchases the first 1,487.5 of a total of 7,332 acres from the Hannahs' "Homestead Trust" in Sandwich and Bourne for the Coonamessett Ranch Company.
1915: Wool merchant George St. Amant from Newton, Massachusetts, buys 66 acres on the southwest side of Coonamessett Pond from the

estate of Joseph Story Fay to initially establish a hunting camp and later a dairy farm called Atamannsit; investment advisor Frederick Leatherbee bought sixty parcels of land totaling 2,936 acres, including several farms, mainly in Hatchville for the Coonamessett Ranch Company.

1916: George Tent buys Wallace Hatch's one-hundred-acre homestead for a chicken farm; Coonamessett Ranch Company incorporates in Massachusetts as a business in agriculture and timber, as well as land sales and development. The directors are Charles Crane and his son Richard Crane, Wilfrid Wheeler and Robert and Frederick Leatherbee.

1917: Coonamessett Ranch Company now owns 11,638 acres of land in Falmouth, Bourne, Sandwich and Mashpee; Wilfrid Wheeler plans the Coonamessett Ranch Company as a colonization scheme to help train and increase the number of farmers on Cape Cod.

1919: Wilfrid Wheeler serves one year as the first commissioner of the reorganized Massachusetts Department of Agriculture; Silas Hatch dies in Hatchville; Wilfrid Wheeler begins a five-year contract to manage the Coonamessett Ranch Company's model farm in Hatchville.

1920: Falmouth population 3,500.

1921: Coonamessett Ranch Company now owns fourteen thousand acres of contiguous land in Upper Cape Cod; estimated total investment by Charles R. Crane is $275,000 for land and model farm; Robert Leatherbee begins to purchase Hatchville land for Brae Burn Farms.

1924: Robert Leatherbee and Frances Anita Leatherbee divorce; Coonamessett Ranch Company directors abandon the model farm and replace it with an inland summer resort; C. Morris Williams starts Crooked Pond Farm on Hatchville Road.

1925: Coonamessett Inn opens as anchor of the new Coonamessett Resort; Wilfrid Wheeler starts Ashumet Farm in Hatchville south of Ashumet Pond; Robert Leatherbee marries Falmouth resident Helen Everhart.

1926: After purchasing George Tent's poultry farm, Robert Leatherbee's land holdings for Brae Burn Farm increase to approximately four hundred acres.

1928: Coonamessett Resort's golf course opens with nine holes. The next year, nine more holes were added, and a new clubhouse was built.

1929: Frederick Leatherbee starts Kensington Dairy Farm on Hatchville Road; Atamannsit Farm's prized Guernsey cattle are sold at auction.

1930: Falmouth population 4,821.

1934: Kensington Farm files for bankruptcy, and its assets and land are sold to Robert Leatherbee; Robert W. Leatherbee, fifty-three, dies in Hatchville.

1935: Coonamessett Ranch Company sells 62 percent of its Upper Cape Cod land to the Commonwealth of Massachusetts for the expansion of Camp Edwards within the Massachusetts Military Reservation.

1936: Robert Leatherbee's real estate is sold; Hjalmar Jensen, Brae Burn Farms's manager, purchases 121 acres from Leatherbee estate to start a dairy/truck farm; Norman Dupee, a Boston wool merchant, purchases a 76.7-acre parcel, including the Lochstead manor house, from the Leatherbee estate for a family compound.

1938: Norman Dupee purchases at foreclosure about 150 acres along the southwest shore of Coonamessett Pond, which includes the former Atamannsit Farm, for summer cottages and later modest homes for his adult children.

1939: Charles R. Crane dies on February 14 at his winter home in Palm Springs, California; John O. Crane and his sister, Frances Anita form a partnership called the Coonamessett Associates to run the Coonamessett Resort.

1940: Falmouth population 6,878.

1941: Coonamessett Associates build twenty-five cottages along both sides of Boxberry Hill Road and construct the Coonamessett Club to provide year-round recreation; Cornelia Workman Smith Crane (wife of Charles R. Crane) dies in Woods Hole.

1942: Dimmocks start Hunky Dory Farm on Hatchville Road.

1949: Coonamessett Club remodeled into an air-conditioned, six-hundred-seat Falmouth Playhouse and leased to producer Richard Aldrich.

1950: Falmouth population 8,662.

1953: Treadway Hotel Company invests in the Coonamessett Resort.

1954: Frances Anita Crane, sixty-six, is killed in automobile crash on August 10.

1956: John O. Crane and Frances Anita Crane's estate trustees sell the Falmouth Playhouse to Sara (Sidney) Gordon.

1957: John O. Crane and Frances Anita Crane's estate trustees sell the golf course, Treadway Inn, the motor lodge and the twenty-five guest cottages to Harvey G. Clauson Sr; Clauson's Inn and Resort opens.

1958: Massachusetts Division of Fisheries and Wildlife purchases from John O. Crane and Frances Anita Crane's estate trustees 1,562.1 acres of woodland and grassland parallel to Route 151 and the Coonamessett airport. The area is named "Frances A. Crane Wildlife Management Area."

1958: Harvey G. Clauson Sr. dies in a house fire, leaving his wife, M. Elizabeth Clauson, and son Donald Clauson in control of Clauson's resort.

1960: Falmouth population 13,037.

1961: Wilfrid Wheeler of Ashumet Farm dies in Hatchville.

1962: Wilfrid Wheeler's heirs sell Ashumet Farm's horticulture plantings (mainly hollies) to local benefactor and holly grower Josiah K. Lilly III, who donates the land to the Massachusetts Audubon Society.

1970: Falmouth population 15,942; Hatchville Post Office is closed by U.S. Postal Service after 112 years.

1972: New Alchemy Institute is started on Hatchville Road by Nancy Jack Todd, John Todd and William McLarney on leased land.

1977: M. Elizabeth Clauson sells Clauson's Resort, including the golf course, to Steps Realty Trust (Stephen Hopkins and Shepard Sumberg) for $1.475 million (the pair planned to develop 250 single-family homes near the golf course and Coonamessett Pond—this did not happen); Steps Realty Trust sells Clauson's golf course, now called Cape Cod Country Club, to John Gilgun, former mayor of Woburn, Massachusetts, and golf professional Ernest Doherty.

1980: Falmouth population 23,640.

1980: Steps Realty Trust enters into Chapter 11 dissolution of its assets.

1981: 216 acres of former Steps Realty Trust resort land transferred to Silent Ridge Trust. Silent Ridge Trust transfers its assets to Oliver Street Corporation.

1985: Hatchville Development Realty Trust (HDRT) files preliminary plans for a 336-acre subdivision called "The Country Club on Coonamessett," with the Falmouth Planning Board on June 12. Two new organizations opposed to HDRT's planned subdivision are formed: Coonamessett Pond Association and The 300 Committee.

1986: On February 4, the Falmouth Planning Board votes to approve the latest preliminary plans of HDRT for a cluster plan of 225 units—31 multifamily and 101 single-family residences—on 359 acres in the heart of historic Hatchville; on April 7, the special town meeting votes 175 to 18 to acquire eight tracts of land totaling 401 acres for $8.3 million (the largest and most expensive of these parcels, the "Coonamessett Pond, Hatchville" tract of 212 acres, is purchased for $3.9 million from the Hatchville Development Realty and from the Oliver Street Corporation. A week later, a referendum to authorize the borrowing for the land acquisition garnered over 70 percent of the vote, and the Coonamessett Reservation was born); Lochstead Association forms, and 69 homes are developed north of Coonamessett Pond on former Dupee land.

1989: Coonamessett Farm is started by Ronald and Dr. Roxanna Smolowitz on Hatchville Road.

1990: Falmouth population 27,960.

1994: Falmouth Playhouse burns down.

TWENTY-FIRST CENTURY

2000: Falmouth population 32,660.

2002: At the southwest corner of Sandwich Road and Route 151 on what Harvey Clauson Jr. called "Clausons Corner," a service station and a convenience store are developed; Coonamessett Reservation expands on Route 151, and land around Little Jenkins Pond is protected by Land Bank purchases of land aggregated by Harvey G. Clauson Jr. and others.

2003: Paul R. Dupee Jr. sells to the Town of Falmouth sixty-two acres at Atamannsit for conservation land and fourteen acres on Sandwich Road for recreation fields.

2007: Hatchville Day celebration, August 18; The 300 Committee purchases the ten-acre River Bend Conservation Area with its silo on Coonamessett River.

2009: The 300 Committee purchases the 6.8-acre David and Carmela Bartolomei Conservation Area on Coonamessett River.

2010: Falmouth population 31,531.

2013: Nickelodeon Cinema is sold to Generation AJW LLC.

NOTES

PREFACE

1. Gunning, Gunning and Wheeler, "Hatchville," 236–53.
2. Wyman, "Wilfrid Wheeler," 65–72.
3. Garrick, "80 Swinging Years," 30–33.
4. Michael Freeman, interview 2008.
5. Brigham, *Cape Cod*, 161.

CHAPTER 1

6. www.hatchfamilyassoc.org/research.htm; Geoffrey, *Suckanesset*.
7. Oldale, *Cape Cod, Martha's Vineyard & Nantucket*.
8. Svenning, "18th Century Deeds."
9. *Falmouth Enterprise*, May 25, 1962.
10. General Court of Massachusetts, April 18, 1857.
11. *Official Register of the United States*, July 1, 1883.
12. "East Falmouth Herring River Company in Falmouth," 1863.
13. Witzell, *Joseph Story Fay*.
14. White, "Portuguese in Falmouth," 2–16.
15. Sears, "Preservation of the Cranberry Bogs."

16. Garrick, "Rocks to Riches in Woods Hole," 64–72; Gaines, "Pacific Guano Company," 11–15.
17. Hoffman, "Allan B. Hatch Remembers," in Smith, *Book of Falmouth*, 257–59.
18. Alexander, "Impact of Tourism," 320–26.

CHAPTER 2

19. www.craneco.com/Media/crane_150.pdf; Saul, *Life and Times of Charles R. Crane*. Details about the Crane family circle in Professor Saul's book do not agree in many cases with those in this book.
20. Lillie, *Woods Hole Marine Biological Laboratory*.

CHAPTER 3

21. http://www.nps.gov/caco/historyculture/shipwrecks.htm.
22. *Boston Daily Globe*, "Cape Cod Has Biggest Farm East of the Mississippi," August 4, 1918, 45.
23. BCRD, October 23, 1917, book 353, page 408.
24. Ibid.
25. "Coonamessett Ranch Company, Articles of Organization."
26. Muir, *Reflections in Bullough's Pond*.
27. Margaret Hopkins Freeman, interview, 2008.
28. "Papers of Richard Teller Crane, III,"
29. Crane family papers.
30. "Papers of Richard Teller Crane, III."

CHAPTER 4

31. http://www.mass.gov/eea/agencies/agr/about/history-generic.html.
32. National Conference on Marketing and Farm Credits, "Marketing Milk in New England," 140–44.
33. *Boston Daily Globe*, "Governor Names New State Boards."
34. "Papers of Richard Teller Crane, III."
35. *Falmouth Enterprise*, "Richard Crane Killed by Own Gun."
36. Gunning, Gunning and Wheeler, "Hatchville."
37. Wallbank, "Fertile Spread at Coonamessett," May 7, 1972.

38. BCRD, book 415, pages 497–98, February 16, 1925.
39. Charles Wheeler. Recorded, transcribed interview, April 17, 1986.
40. *Cape Cod Standard Times*, "Hatchville Man Has Recorded Rainfall," November 30, 1953.
41. Wyman, "Wilfrid Wheeler," 65-72.
42. Garrick, "Banking on Beach Plums," 2012.
43. Garrick, "The Man Who Planted Ashumet Holly," 2005.
44. *Falmouth Enterprise*, "Tree Planting Committee Report."
45. "Changing Nature of Ashumet," *Mass Audubon*, 2010; Mills, "Ashumet Holly Reservation," 1975.

CHAPTER 5

46. BCRD, January 25, 1921, book 375, page 567.
47. *Waukegan Sun*, "Divorce Given to Woodrow's Campaign Mgr."
48. Helen Everhart and her younger sister Mary were the stepdaughters of Falmouth physician Dr. Lombard C. Jones and his wife, Nina.
49. Until he was of legal age to enlist in the U.S. Navy, James Leatherbee lived with his mother, Helen, and his stepfather in Hingham, Massachusetts. He later married and had one child, Richard Leatherbee. James died in 2006 at age seventy-nine.
50. *New York Times*, "Jan Masaryk Weds Mrs. F.C. Leatherbee," December 24, 1924. Charles R. Crane was in Egypt at the time of the hasty wedding.
51. Joseph Hoffenberg, "The University Players," in Smith, *Woods Hole Reflections*, 275–77.
52. Houghton, *Entrances & Exits*.
53. BCRD, June 9, 1922, book 388, page 132.
54. Ibid., August 31, 1926, book 459, page 343.
55. Leatherbee family archives.
56. BCRD, March 29, 1929, book 464, page 151.
57. Ibid., March 25, 1929, book 463, page 429.
58. *Falmouth Enterprise*, "R.W. Leatherbee Funeral Today."
59. In 1924, John Wray moved his dairy business from the Fay Farm on Elm Road in Falmouth to the site of the old Hatchville schoolhouse at Ashumet and Boxberry Hill Roads before working for the Coonamessett Ranch Company and Brae Burn Farms.
60. *Falmouth Enterprise*, "Robert W. Leatherbee."

61. Ibid., "R.W. Leatherbee Wills $130,000."
62. BCRD, Hunt: May 7, 1937, book 523, page 564; Temple: May 5, 1937, book 527, page 20; Ellis: October 13, 1936, book 520, page 360.
63. Ibid., September 17, 1936, book 522, page 340.
64. Barnstable County Land Court, A4 8838-C, February 1, 1937.
65. BCRD, October 17, 1938, book 544, page 417.
66. Ibid., October 7, 1937, book 530, page 536.
67. Ibid., October 19, 1937, book 529, page 389; March 17, 1939, book 549, page 574.

CHAPTER 6

68. Garrick, "80 Swinging Years."
69. *Falmouth Enterprise*, "State Ready to Pay Coonamessett Ranch $50,000."

CHAPTER 7

70. Crane family papers.
71. Sylvia E. Crane memoirs, 2003.
72. *Falmouth Enterprise*, "John Crane Opens a Restaurant."
73. Ibid., "Mrs. Crane Becomes Falmouth Voter."
74. Ibid., "Cranes Win Zoning Change."
75. Ibid., "Permit to Operate Hatchville Airport Denied by Appeal Board."
76. Ibid., "Airshow Audience Stunned at Death of Parachutist."
77. Ibid., "New Playhouse is Ready for Tallulah Tonight."
78. Bankhead, *Tallulah*, 246–47.
79. *Falmouth Enterprise*, "New Yorker Buys Mrs. Crane's House."
80. Ibid., "Champagne Poured at Coonamessett but Sale and Lease Both Denied."
81. Bedford, "Edna Harris."
82. This was Ms. Bankhead's first stage role in three years. In *Dear Charles*, she played a mother who explains to her three children that they all have different fathers. Monday, August 9, was the opening for this play. The Playhouse was closed on Sunday, and all seats were reserved. Prices were between $1.20 and $3.60 for most nights and matinees with a top price of $4.20 on Saturday evening.

83. *Falmouth Enterprise*, "Triumph Marked Her Cosmopolitan Life."
84. Ibid., "Airman Appeals Crash Decision; Suit Is Filed."
85. Ibid., "Sons to Share Crane Estate"; Barnstable Probate Court, Case no. 34363.

CHAPTER 8

86. Carol Baker, interview, 2008.
87. Margaret Freeman, interview, 2008.
88. Ibid.
89. Nancy Williams Chute, interview, 2008.
90. *Falmouth Enterprise*, "War Prisoners' Life in Edwards Stockade." One solution to overcome the labor shortage during the war was to bring prisoners, including approximately 150,000 from North Africa, back to the United States to use their labor to help in particular segments of the economy. Camp Edwards on Cape Cod housed 2,000 prisoners of the German Afrika Korps. Prisoners were paid for their work on and off the base; Krammer, *Nazi Prisoners of War in America*, 268–72. Camp Edwards was one of 142 major German POW facilities in the USA, ultimately housing nearly 375,000 men.
91. Fred Jensen, interview, 2008.
92. Ibid.
93. Bertram Tomlinson was the Barnstable County extension agent from 1924 to 1959. He educated farmers on new techniques, especially the use of chemicals as fertilizers and the application of herbicides. Tomlinson worked closely with the cranberry and strawberry growers, and during World War II, he rallied the public to help harvest the valuable strawberry crop. Postwar, he helped organize the Cape Cod Beach Plum Growers Association.
94. In 1947, the congregation celebrated the EEMH's 150th anniversary. Events included a card party, dance and a strawberry fair leading up to a final celebration at the church and a candlelight supper at the parsonage. Over the next several decades, the church fell into disrepair. For example, in the early 1960s, services could no longer be held until the roof and windows were repaired and a new ceiling installed. Funds were raised, and the East End Congregational Religious Society was organized as a

corporation. By the winter of 1966, funds were available to restore the old buildings. The congregation held on for a few more years.

In the late 1970s, the EEMH was home to an interdenominational congregation led by Reverend Eugene D. McKenney. The 1978 "Hatchville Happening Day" at the EEMH was celebrated with musical performances, pony rides and food.

In 1981, the Falmouth Jewish Congregation (FJC) was formed. At first, the congregation met in people's homes and then at the chapel at St. Barnabas Church in Falmouth. The land and buildings of the EEMH were transferred to the FJC on November 23, 1982. In 1992, the FJC built a center for religious, education and social functions on former Dupee land.

95. Coonamessett Farm in Hatchville (http://www.coonamessettfarm.com/) is a twenty-acre farming and research enterprise organized in three divisions: agricultural production, research and consulting.

96. Todd, *A Safe and Sustainable World*, 138.

97. The Green Center Inc. (http://www.thegreencenter.net), a nonprofit educational institute, is the custodian and distributor of publications of New Alchemy's ecological research conducted from 1971 to 1991. The Green Center focuses on wastewater treatment, especially eco-toilets.

98. Cape and Islands Self-Reliance Coalition is a spin-off of NAI that promotes environmentally sound energy technologies (http://www.reliance.org/). The Coonamessett Farm Foundation supports scientific research in agricultural production and aquacultural fisheries (http://www.coonamessettfarmfoundation.org/).

CHAPTER 9

99. *Falmouth Enterprise*, "John Crane Opens a Restaurant."

100. Elizabeth Clauson, e-mail to author, December 12, 2012. Elizabeth Clauson, who worked at the Falmouth Playhouse, recalled: 'Sidney was very outgoing and called everyone 'darling.' The character 'Auntie Mame' easily could have been based on Mrs. Gordon"; Lawson, *Theater on Cape Cod*.

101. BCRD, March 20, 1970, book 1466, page 380; Foreclosure: book 1526, page 750.

102. Driscoll, "Ralph Miller Name Notorious in Falmouth"; "Up in Flames: 3 Theaters Owned by Ralph Miller Meet Fiery End."

103. Falmouth Town Report, 1957.
104. Falmouth Town Meeting, article 66, March 10, 1960.
105. DFW's *General Management Plan for Crane Wildlife Management Area*, 1962.
106. George Clauson, interview, 2011.
107. *Falmouth Enterprise*, "Death Ends Long and Successful Business Career."
108. Garrick, "80 Swinging Years,"
109. *Lopez v. United States*, 1963.
110. Harvey G. Clauson Jr. and his first wife settled in West Falmouth, Massachusetts, on the thirty-acre farm formerly owned by Reuben Landers. He opened his first automobile dealership in Falmouth in 1958 and sold his last one, a Chevrolet and Oldsmobile sales and service center, in 1987. Mr. Clauson also developed residential properties in West Falmouth.

CHAPTER 10

111. *Saving Losing Cape Cod*, download at http://www.whrc.org/mapping/capecod/index.html.
112. Carroll, *Boston Globe*, February 13, 1992, 56; *United States v. Patricia C. Harrison*, 78 F3d 577, http://openjurist.org/78/f3d/577/95/2009.
113. *Falmouth Enterprise*, "250 House Lots Are Planned for Clauson Tract."
114. BCRD, April 27, 1981, book 3275, page 174.
115. Phil Friel was a professional golfer and close friend of Ernest Doherty. Friel is a member of the New England PGA Hall of Fame and was a NE PGA champion and a golf course designer.
116. Lowell, "Report to the Tricentennial Commission," 1986.
117. *Falmouth Enterprise*, "Town Meeting Approves Land Acquisition Program by 175 Votes to 18."; ibid., "Voters Affirm Spending $8.3 million for Open Space."
118. Ibid., "Size of Development on Coonamessett Pond Concerns Planners."
119. BCRD, May 20, 1986, Doc. No. 39444.
120. Betsy Gladfelter, interview, 2008.
121. Barnstable County Land Court plan 12349-11, April 8, 1997.
122. Garrick, "80 Swinging Years."
123. BCRD, Lot 182: February 7, 2002, book 14877, page 89; Probate Court, Docket no. 65968. M. Elizabeth Clauson, the mother of Donald, Arlene Robbins and Harvey, died on July 26, 1983, in Falmouth at age eighty-four. Donald Clauson was her primary heir.

124. Town of Falmouth Land Bank 1998–2007; Falmouth Town Report, 2004; Falmouth Town Meeting, article 6, April 3, 2002.
125. 300 Committee, "Coonamesset River Gateway," http://www.300committee.org/land.htm.

CODA

126. Elias Lieberman, interview, 2008.
127. Henry Dick, interview, 2008.
128. Louv, *Nature Principle*; Louv, *Last Child in the Woods.*
129. Jay Zavala, interview, 2009.

BIBLIOGRAPHY

Alexander, Lewis M. "The Impact of Tourism on the Economy of Cape Cod, Massachusetts," *Economic Geography* 29 (1953): 320–26.

Atlas of Barnstable County, Massachusetts. Boston: Walker Lithograph & Publishing Co., 1910.

Baker, Carol. Interview by the author, November 15, 2008.

Bankhead, Tallulah. *Tallulah, My Autobiography*. New York: Harper Brothers, 1952.

Barnstable County Land Court A4 8838-C, February 1, 1937.

Barnstable County Land Court plan 12349-11, April 8, 1997.

Barnstable County Probate Court. Case no. 34363. Robert W. Leatherbee, August 1954.

———. Docket no. 65968. M. Elizabeth Clauson, July 26, 2003.

Barnstable County Registry of Deeds (hereafter BCRD).

Bedford, Ethelda. "Edna Harris, Innkeeper Extraordinary." *Yankee Magazine* (1957).

Boston Daily Globe. "Cape Cod Has Biggest Farm East of the Mississippi." August 4, 1918, 45.

———. "Governor Names New State Boards, Many Well-Known Men Lose Places." November 25, 1919, 1.

Brigham, A.P. *Cape Cod and the Old Colony*. New York and London: G.P. Putnam's Sons, 1920: 161.

Cape Cod and All the Pilgrim Land 6, no. 4 (June 1922).

Cape Cod Standard Times. "Hatchville Man Has Recorded Rainfall for nearly 30 Years." November 30, 1953.

Carroll, Matt. "Falmouth Woods Partners Plead Innocent to Fraud." *Boston Globe*, February 13, 1992, 56.

"The Changing Nature of Ashumet." *Mass Audubon*, 2010.

Chute, Nancy Williams. Interview by the author, October 2, 2008.

Clauson, George. Interview by author, March 26, 2011.

"Coonamessett Ranch Company, Articles of Organization." Certificate of incorporation. Issued September 16, 1916. Record book 339, page 53.

Crane family papers, 1875–1980. Bakhmeteff Archive of Russian and East European History and Culture, Rare Books and Manuscripts Library, Columbia University.

Crane, Sylvia E. *Memoirs: 1918–2002*. Princeton, NJ: privately printed by Sylvia E. Crane, 2003.

Deyo, Simeon L., ed. *History of Barnstable County, Massachusetts*. New York: H.W. Blake & Co, 1890.

Dick, Henry. Interview by author, November 11, 2008.

Driscoll, Kathi Scrizzi. "Ralph Miller Name Notorious in Falmouth." Cape Cod online.com, October 25, 2009.

———. "Up in Flames: 3 Theaters Owned by Ralph Miller Meet Fiery End." Cape Cod online.com (accessed February 24, 2013).

Falmouth Enterprise. "Airman Appeals Crash Decision; Suit Is Filed." November 20, 1954, 6A.

———. "Airshow Audience Stunned at Death of Parachutist." November 5, 1948, 2.

———. "Bill to Create Town of Nobska Was Secession Threat in 1907." May 25, 1962, Section F, 1.

———. "Champagne Poured at Coonamessett but Sale and Lease Both Denied." October 4, 1946, 1.

———. "Cranes Win Zoning Change." February 13, 1942, 8.

———. "Death Ends Long and Successful Business Career." May 6, 1958, 1.

———. "John Crane Opens a Restaurant on Rome's Ancient Appian Way." January 24, 1968.

———. "Mrs. Crane Becomes Falmouth Voter." January 30, 1942, 9.

———. "New Yorker Buys Mrs. Crane's House." October 28, 1949.

———. "Permit to Operate Hatchville Airport Denied by Appeal Board." November 1, 1946, 1.

———. "Richard Crane Killed by Own Gun While Hunting on Virginia Estate." October 7, 1938, 2.

———. "Robert W. Leatherbee." Obituary. August 9, 1934.

———. "R.W. Leatherbee Funeral Today." August 9, 1934, 1.
———. "Sons to Share Crane Estate." August 20, 1954.
———. "State Ready to Pay Coonamessett Ranch $50,000 for Guard Camp Land." September 26, 1935, 1.
———. "Town Meeting Approves Land Acquisition Program by 175 Votes to 18." April 9, 1986, 1.
———. "Tree Planting Committee Report Is Survey of Two Years' Work." January 24, 1947, 12.
———. "Triumph Marked Her Cosmopolitan Life." August 13, 1954, 1.
———. "250 House Lots Are Planned for Clauson Tract." April 22, 1977.
———. "Voters Affirm Spending $8.3 million for Open Space." April 16, 1986, 1.
———. "War Prisoners' Life in Edwards Stockade." July 14, 1944, 3.
Falmouth Land Bank. "Open Space Acquisition by Town of Falmouth 1999–2007." March 2008.
Falmouth Town Meeting. Article 6, April 3, 2002.
———. Article 66, March 10, 1960.
Falmouth Town Report, 1957, 2004.
Freeman, Margaret Hopkins. Interview by the author, November 22, 2008.
Freeman, Michael. Interview by the author, November 22, 2008.
Gaines, Jennifer Stone. "Pacific Guano Company." *Spritsail* 21 (2007).
Garrick, Les. "Banking on Beach Plums," *Gastronomica: The Journal of Food and Culture* (2012): 21-30.
———. "80 Swinging Years: Making and Remaking Coonamessett Golf." *InSeason* supplement to *Falmouth Enterprise*, May 2011.
———. "The Man Who Planted Ashumet Holly." Published online by massaudubon.org, 2005.
———. "Rocks to Riches in Woods Hole." *Cape Cod Life* (2006).
General Court of Massachusetts. "An Act to Authorize Silas Hatch and Others to Construct a Cart-way in the Town of Falmouth." April 18 1857, Chapter 70. Accessed via Google books, October 9, 2012.
Geoffrey, Theodate [Dorothy Godfrey Wayman]. *Suckanesset: A History of Falmouth Massachusetts*. Reprinted by the Falmouth Historical Society, 1992.
Gladfelter, Elizabeth (Betsy). Interview by author, October 30, 2008.
Houghton, Norris. *Entrances & Exits: A Life in and out of the Theatre*. Montclair, NJ: Limelight Editions, 1991.
Jensen, Fred. Interview by the author, October 6, 2008.
Krammer, Arnold. *Nazi Prisoners of War in America*. New York: Stein and Day, 1979.

Lawson, Evelyn. *Theater on Cape Cod.* Yarmouth Port, MA: Parnassus Imprints, 1969.

Leatherbee family archives.

Lieberman, Elias. Interview by author, November 17, 2008.

Lillie, Frank R. *The Woods Hole Marine Biological Laboratory*. Chicago: University of Chicago Press, 1944.

Lopez v. United States. http://supreme.justia.com/cases/federal/us/373/427/case.html.

Louv, Richard. *Last Child in the Woods: Saving Our Children from Nature-Deficit Disorder*. Chapel Hill, NC: Algonquin Books of Chapel Hill, 2008.

———. *The Nature Principle: Human Restoration and the End of Nature-Deficit Disorder*. Chapel Hill, NC: Algonquin Books of Chapel Hill, 2011.

Lowell, Vicky. "Report to the Tricentennial Commission," 1986.

Massachusetts Division of Fish and Wildlife. *General Management Plan for Crane Wildlife Management Area*. 1962.

Mills, Elwood. "Ashumet Holly Reservation." Massachusetts Audubon Newsletter 14 (1975).

Muir, Diana. *Reflections in Bullough's Pond, Economy and Ecosystem in New England.* Hanover, MA, and London: University Press of New England, 2000.

National Conference on Marketing and Farm Credits. "Marketing Milk in New England." Chicago: National Conference on Marketing and Farm Credits and National Council of Farmers' Cooperative Assocations, 1915. Accessed as Google e-book.

New York Times, "Jan Masaryk Weds Mrs. F.C. Leatherbee." December 24, 1924.

Official Register of the United States, the Post Office Department and the Postal Service. Vol. 2. July 1, 1883 (accessed via Google Books, October 9, 2012).

Oldale, Robert N. *Cape Cod, Martha's Vineyard & Nantucket: The Geologic Story.* Yarmouth Port, MA: On Cape Publications, 2001.

"Papers of Richard Teller Crane, III." Special Collections, Lauinger Library of Georgetown University, Washington, D.C.

Saul, Norman E. *The Life and Times of Charles R. Crane, 1858–1939: American Businessman, Philanthropist, and a Founder of Russian Studies in America.* Lanham, MD: Lexington Books, 2012.

Sears, Ann. "Preservation of the Cranberry Bogs of Cape Cod as Rural Historic Landscapes." Master of Arts thesis, Goucher College, 2009.

Smith, Mary Lou, ed. *The Book of Falmouth: A Tricentennial Celebration; 1686–1986.* Falmouth, MA: Falmouth Historical Commission, 1986.

———. *Woods Hole Reflections.* Woods Hole, MA: Woods Hole Historical Collection, 1983.

Svenning, Dorothy. "18th Century Deeds." *Spritsail* 15 (2001).

The 300 Committee. "Coonamessett River Gateway." http://www.300committee.org/land.htm.

Todd, Nancy Jack. *A Safe and Sustainable World.* Washington, D.C.: Island Press, 2005.

United States v. Patricia C. Harrison, 78 F3d 577. http://openjurist.org/78/f3d/577/95/2009.

Wallbank, Lucy M. "Fertile Spread at Coonamessett." *Sunday Standard-Times*, May 7, 1972.

Waukegan Sun. "Divorce Given to Woodrow's Campaign Mgr." January 15, 1924.

Wheeler, Charles. Recorded, transcribed interview, April 17, 1986. Woods Hole Public Library.

White, Lewis. "The Portuguese in Falmouth: 1870 to 1930." *Spritsail* 27 (2013).

Witzell, Susan Fletcher. *Joseph Story Fay: The Life of a Modest Man.* Printed privately, 2013.

Woods Hole Research Center, "Saving Losing Cape Cod," Download at http://www.whrc.org/mapping/capecod/index.html.

www.craneco.com/Media/crane_150.pdf.

www.hatchfamilyassoc.org/research.htm.

www.mass.gov/eea/agencies/agr/about/history-generic.html.

www.nps.gov/caco/historyculture/shipwrecks.htm.

Wyman, Donald. "Wilfrid Wheeler: The Holly Man." *Arnoldia* 14 (1954).

Zavala, Jay. Interview by author, January 12, 2009.

INDEX

C

F

G

H

ABOUT THE AUTHOR

Les Garrick shares his curiosity about the world with anyone willing to read his work and view his photographs. Are you that person? He approaches historical studies without too many biases by asking a simple question: How did this or that come to be this way? You never know where it will lead. From Woods Hole on Cape Cod, he has explored local history and historical figures. His latest effort was a privately published study, "Ward Mayhew Parker: A 19th Century Woods Hole and New Bedford Seaman and Businessman," in 2012. In the interim, he indulged one of his other passions, discovering the many values and virtues of the beach plum, an iconic native seaside shrub. His latest article about the beach plum appeared in the prestigious journal *Gastronomica: The Journal of Food and Culture* in 2012. Les also promotes local food security by supporting farmers' markets and helping people start and maintain vegetable gardens. In his early years, he actively worked to protect crocodilians and their habitats and carried out pioneering research of their social behavior.